I0494771

MACHIAN RELATIVITY

The New Theory That No Scientist Can Disprove

By

Glyn Phillips

Machian Relativity

Copyright © 2012 Glyn Phillips

All rights reserved

ISBN: 1470057212
ISBN-13: 9781470057213

There are really four dimensions, three which we call the three planes of Space, and a fourth, Time.

H. G. Wells, The Time Machine (1895)

PREFACE

This book offers the reader something different from the usual publications on relativity. Most other books on the subject are written by scientists and authors working within the framework of Establishment Science. Consequently, these publications all provide the usual standard explanations of relativity. One aim of this book is to prove that many of those explanations are false. Such science books are published and promoted because their authors (usually professors and researchers) are considered "experts", although this restriction doesn't seem to apply to writers of other subjects. But non-academics (including former graduates) have to use self-publishing, because they are not "experts".

Neither can outsiders get their ideas considered by scientific organisations as such ideas have not been "peer reviewed". Einstein's early papers were published with only informal peer review when he was a patents clerk. But as the practice became more formalised he did everything he could to avoid it. When a journal peer reviewed his paper "Do Gravitational Waves Exist?" and found an error, he simply sent his paper to another journal that didn't have peer review. The paper was then published with the error. Peer review now involves academics only scrutinising each other's papers. University professors and journals do not consider the ideas of outsiders, regardless of their qualifications. And that includes patents clerks. So ironically, Einstein's early papers would never have seen the light of day in the current system. There are several ways in which alternative ideas can be kicked into the long grass, and the ideas of outsiders are easily dismissed as being those of amateurs or cranks.

We also need to consider a certain fact about experts. If experts were always right, then everything that could be discovered would have been discovered thousands of years ago, and there would be nothing for today's scientists to do. The nature of many a theory is that often some part of it turns out to be incomplete or false. For nearly two thousand years, experts believed in Aristotle's theories of motion, until they were refuted by Galileo. And there's no reason why the same might not happen someday to one of today's theories. The length of time a theory has been around, the social standing of its creator, and how many supporters it has, are irrelevant. Ernest Rutherford discovered the atomic nucleus, but believed using its energy as a power source was "moonshine". Even Einstein could get it wrong sometimes. His theory of General Relativity predicted an expanding universe, but he didn't like the result so he fudged the calculations to predict a static universe. And if he was wrong about these things, then he might also have been wrong about Special Relativity.

Experts have also dismissed the ideas of others, only to embrace them later. Alfred Wegener's theory of continental drift was ridiculed by the science community for decades, and Frank Whittle's invention of the jet engine was initially dismissed by an expert as being "too simple". There is no reason to think the situation today is any different. But people are generally better informed now than at any previous time in history. And most people who graduate from universities don't stay in such places, but this doesn't mean they never have their own ideas. So it would be wrong to casually dismiss someone's idea, just because professional scientists had not thought of it earlier. An idea should be judged solely on its own merits, and not the professional status of its creator.

What is this book about? Firstly, I have developed arguments that definitively prove that Special Relativity is wrong. Secondly, I have developed an alternative to Special Relativity, which I call Machian Relativity. I am its sole creator. The argument is that Special Relativity and conventional "explanations" of the Twins Paradox are unsatisfactory for a variety of reasons:

1. Acceleration effects *per-se* cannot explain the asymmetric ageing result of the twins. Applied forces cause non-inertial effects only while they are acting. Likewise, they can only cause asymmetric time dilation while they act, so they cannot compensate for the paradoxes of reciprocal dilation when the travelling twin moved inertially.
2. Scientists propose various "explanations" based on Time Delays or General Relativity to circumvent the above

problem. These introduce faster ageing of the earth twin relative to the space twin, but they differ in where this occurs in the journey. And none of them are used in practical applications (particle accelerators or the Hafele-Keating Experiment) which instead require continuous dilation asymmetry during motion.

3. There is an attempt to hide the paradoxical nature of reciprocal clock rates by claiming that inertially moving clocks cannot be compared. This is false.

4. The asymmetric result predicted by these "explanations" (fudge factors) is identical to the simpler assumption that the travelling twin has instead aged asymmetrically for every part of his journey, so this is probably what is actually happening.

It is obvious that these "explanations" have been concocted by scientists to save Special Relativity from its illogical consequences. Physicists have never attempted to define one as the definitive version; they cannot all be correct as their details are different and have not been experimentally verified. These "explanations" retain Einstein's reciprocal dilation, but there is no specific experimental evidence for it. All the experimental results on time dilation are asymmetric, and scientists have never proposed to look for the symmetric aspect. Consider the Hafele-Keating Experiment: the earth clock and aircraft clock both undergo asymmetric dilation relative to a non-rotating frame, during their motion. There is no reciprocal dilation effect between them, only an asymmetric change in their relative rates. And the fact that earth observers need to include the dilation of their own clocks disproves Einstein's interpretation of the Michelson-Morley null result that earth observers are "at rest".

Machian Relativity is better because:

1. The equations derive the asymmetry from First Principles. This is done by including the relative effects of mass in the equations. This leads to the asymmetry being relative to the average momentum of matter in the universe, thereby uniting relativity with Mach's Principle.

2. There is no need to postulate the existence of the Aether, as absolute motion and absolute rest are self-generating from the average momentum of all matter.

3. The equations are inherently asymmetric for both inertial motion and acceleration. This is simpler than conventional physics, which has to alternate between symmetric and asymmetric dilation depending on the state of motion.

4. The asymmetric ages of the twins (or clocks) on reunion is due to asymmetric ageing during the journey itself. There can be no objection to this, as it is consistent with the asymmetric dilation of moving clocks in the Hafele-Keating Experiment. Therefore, there is no need for supplementary "explanations" to save the theory.

Machian Relativity retains many of the features of Special Relativity, including the Principle of Constancy of the Speed of Light and $E=mc^2$. The major aspect I have changed is the Relativity Postulate, to be compatible with absolute motion. This means that the Laws of Physics are "the same" because relativistic changes (such as Time Dilation) in all reference frames are asymmetric and relative to the average momentum of matter in the Universe – the "Fixed Stars". Thus unlike in Special Relativity where covariance is assumed to be due to inertial motion being purely relative, the covariance of the space-time equations in Machian Relativity is an *apparent* effect only. I have also modified Mach's original interpretation of his idea, so that non-inertial effects are not due to acceleration relative to the "Fixed Stars" per-se but to the average momentum of all matter.

In this book I have defined $\gamma = \sqrt{(1 - v^2/c^2)}$ so that $t' = \gamma t$, etc. Standard notation instead has $\gamma = 1 / \sqrt{(1 - v^2/c^2)}$ so that $t' = t / \gamma$, etc. Both versions give $t' = t \sqrt{(1 - v^2/c^2)}$ so there is no quantitative difference. I also define $t_B = \gamma t_\alpha$ where; B has absolute motion and α is at absolute rest. Also, conventional science usually changes notation for relativistic mass and energy, e.g. $m = \gamma m_0$ (instead of $m' = \gamma m$).

Glyn Phillips

CONTENTS

ALPHA

INTRODUCTION

Einstein's Theory of Special Relativity is wrong. But I have rectified it by developing a new theory of relativity. And the beauty of this new theory is that it uses Special Relativity as a starting point, together with phenomena which are inherent in Einstein's equations which have hitherto not been utilized. My method can be regarded as an extension of Einstein's theory, but an extremely radical one. This is; the unification of relativity theory with the ideas of Ernst Mach (something which Einstein himself tried, but failed, to achieve). In developing this new theory, I have made other discoveries regarding the nature of space and time, particularly relating to systems undergoing linear acceleration and rotation. I have also integrated inertia and electrodynamics within this new framework.

A Brief History of Relativity

Other scientists were working on relativity before Einstein published his theories. George FitzGerald proposed relativistic length contraction in 1889 to explain the null result of the Michelson-Morey experiment. The contraction of the apparatus due to the earth's motion through the Luminiferous Aether countered the "Aether wind" effect, producing a null result. Lorentz published his "Electromagnetic Phenomena" in 1904. For clarity I shall call this Aether Relativity (AeR). Einstein published his "Electrodynamics" in 1905. This later became known as Special Relativity (SR). Both theories existed together until Einstein developed his Theory of General Relativity, after which Lorentz's theory was abandoned by the scientific community in favour of Special Relativity. There were several reasons for this.

Firstly, the discovery of the photoelectric effect showed that light existed as particulate photons. Therefore, the idea of photons existing in a void supplanted the earlier idea of light as continuous waves in Maxwell's Luminiferous Aether (a universal substance that filled the void of space). Also, this particulate view of light was consistent with the contemporary discovery that matter was composed of atoms. So, if the Aether was not needed to explain light, then presumably it was not needed to explain relativity either, so Einstein's Theory of Special Relativity (which did not include the Aether) was preferred over Lorentz's theory.

Secondly, Einstein derived effects such as Length Contraction and Time Dilation from first principles by explicitly assuming that the speed of light was constant between relatively moving inertial systems. In contrast, Lorentz initially assumed the contraction and dilation equations (e.g. the length contraction can be inferred from the null result of the Michelson-Morley Experiment), which were physical changes relative to Maxwell's Luminiferous Aether. The speed of light in a vacuum is constant in Maxwell's Equations and Lorentz showed how these changes (in length and time) preserved the form of these equations in frames moving (absolutely) relative to the Aether (implicitly preserving speed of light constancy, which appears as a term in Maxwell's Equations).

Thirdly, Special Relativity was believed to be a special case of General Relativity (GR). Einstein developed his Theory of General Relativity in 1915 to deal with accelerated motion, and was able to explain the perihelion shift in the planet Mercury's orbit; something which Newtonian Gravity could not do. His first theory, which dealt only with inertial motion, was then renamed Special Relativity. Finally, in 1919, experiments during a solar eclipse showed the deflection of starlight by the sun more closely matched the predictions of General Relativity than those of Newtonian Gravity. With the success of General Relativity, and Special Relativity being absorbed into it, there was no need for Maxwell's Luminiferous Aether in this new paradigm.

The Current Situation

The opinions of the scientific community on these matters have remained unchanged ever since. However, Lorentz's theory has one feature worthy of consideration - <u>it gives a far simpler explanation</u> of the Twins Paradox than Special Relativity does. This is because the asymmetric age of the travelling twin (or clock) on return to the earth is due to a physical slowing of his time (asymmetric ageing) *during every part of the journey, for both accelerated and inertial motions.* This is a perfect example of cause and effect – the asymmetric end result in ages (or clock readings) occurs because of a previous asymmetry in ageing (or clock rates) during high speed motion. This fact can be verified by placing synchronised clocks (at rest relative to the earth) along the path of the travelling clock (Fig.1), to determine its rate while it is actually moving (which can be done for different parts of the journey, both accelerated and inertial).

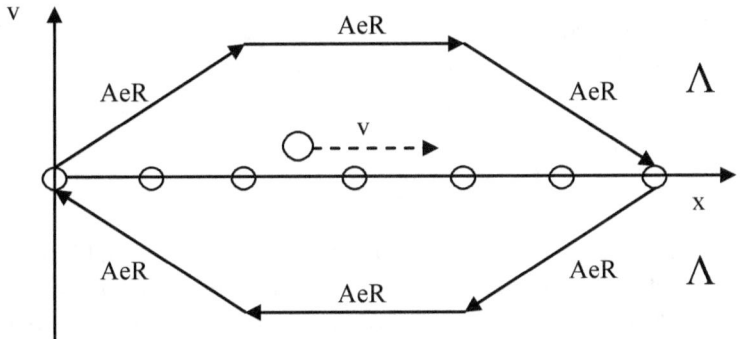

Fig.1

For both inertial and accelerated motion, the space-time asymmetry ($t' = \gamma\, t$, $t = t' / \gamma$) of the moving clock is assumed to be due to motion relative to the "Luminiferous Aether" (Λ), which fills all space and defines a state of universal (absolute) rest.

In contrast, Special Relativity is *incapable of providing a proper explanation of this paradox.* It is stuck with the problem of having symmetric time dilation during inertial motion, and asymmetric dilation during acceleration. The various "explanations" offered by scientists and authors in books to explain the overall asymmetric result when the twins reunite all differ from each other. There is no unanimity or definitive version. Neither is there any agreement on how asymmetry arises during acceleration itself – in the Twins Paradox, General Relativity (GR) is supposedly required, although in practical applications (e.g. the Hafele–Keating Experiment) physicists use Special Relativity in an arbitrarily asymmetric way ("asymmetric Special Relativity", ASR):

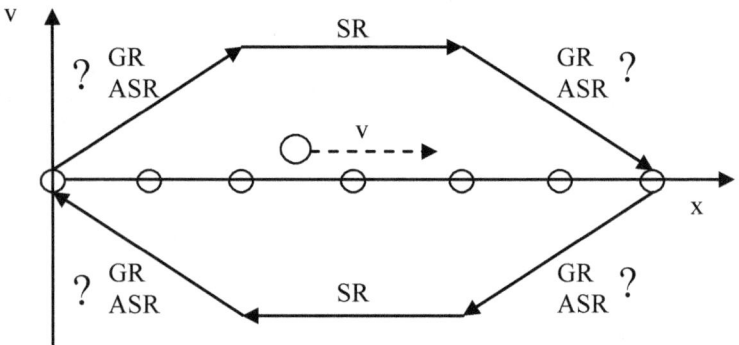

<u>Fig.2</u>

So, for different parts of the journey (Fig.2), Conventional Science yo-yos between symmetric clock rates (Special Relativity) during the inertial motions ($t' = \gamma\, t$, $t = \gamma\, t'$), and asymmetric clock rates (General Relativity) during the accelerations ($t' = \gamma\, t$, $t = t' / \gamma$). Then (somehow), these have to turn into an overall asymmetric result when the clock (or twin) returns to the earth.

The Twins (or Clock) Paradox

The Twins Paradox goes like this. According to Special Relativity, relatively moving inertial clocks work slower than each other (because there is no such thing as absolute motion). This is a reciprocal slowing of time. It also applies to other processes (such as ageing). Now, if one twin stays on earth while the other leaves the earth and goes on a high speed journey, then each twin moves relative to the other. So when the travelling twin returns, Special Relativity means they will be younger than each other (because of their relative motion). This is the Paradox.

Conventional Science then says "the travelling twin has experienced applied forces, and this creates an asymmetry, because he is non-inertial". But the accelerations by applied forces occur for small parts of the journey only, yet they are expected to make the whole journey appear asymmetric.

To circumvent this problem, physicists have proposed a variety of "explanations". One claims that *General Relativity* is required "for a proper analysis" while another claims that General Relativity is not needed, because *Time Delays* provide the answer instead [General Relativity is also used to (supposedly) remove the idea of absolute motion during acceleration by applied forces; but unlike Special Relativity, the time dilation remains asymmetric].

Consider also the asymmetric time dilation of sub-atomic particles during continuous acceleration. How acceleration causes asymmetry in the symmetric equations of Special Relativity is not properly derived from First Principles, but is merely assumed - by *arbitrarily inverting the gamma factor in the equations for non-inertial observers* <u>or</u> *by only using the time dilation equation for inertial observers and algebraically rearranging it.* The experimental results are then said to "prove" Special Relativity. But unlike in the Twins Paradox, no use of time delays or General Relativity is made, even though the motions were non-uniform.

Standard explanations in books also deny that the rate of an inertially moving clock can be directly measured because it needs to "return to the earth for a direct comparison", thus "making it non-inertial". There is also the problem that the equations of Special Relativity, due to their covariance, violate the rules of algebra. So, authors and scientists claim that the equations are "not physical" because they describe how a moving system "appears" to a stationary one, but this is false.

Machian Relativity

It is because of these problems that I developed the Theory of Machian Relativity (MR). This theory is identical to Special Relativity in some respects. Firstly, covariant Lorentz Equations for two relatively moving inertial systems are derived from First Principles (by assuming the speed of light to be constant for both systems). Secondly, the equations for time dilation and length contraction (also known as FitzGerald or Lorentz contraction) are derived from the Lorentz Equations (but in an asymmetric, logically consistent way).

During both inertial and accelerated motion (Fig. 3), there is space-time asymmetry due to the influence of the "Fixed Stars" (Π). This happens despite the symmetry of inertia (Newton's First Law) when systems (clocks) have different constant velocities relative to the *Machian Plenum of the External Universe* (Π). So, asymmetric time dilations for all parts of the journey explain the asymmetric result on reunion. THERE IS NO NEED FOR SUPPLEMENTARY PROCESSES OR "EXPLANATIONS".

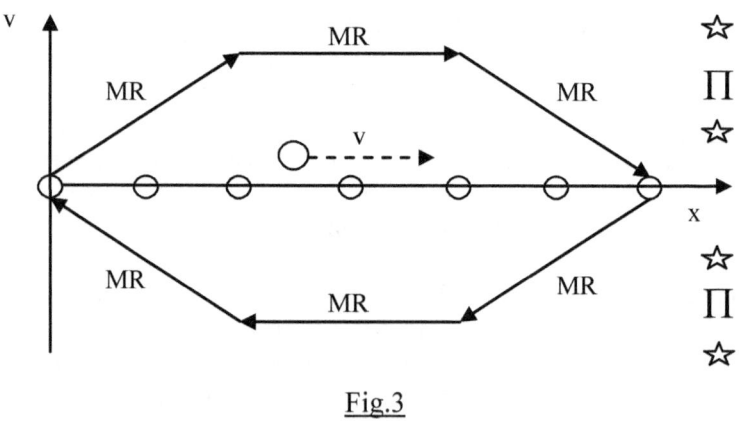

Fig.3

In other respects too, the new theory is different from Einstein's. Unlike Special Relativity, the Relative Masses of systems have an influence on each other, and this is incorporated into the theory. This, together with relative motion, means that the space-time of

6

any system physically changes due to its motion relative to the average momentum of all matter in the universe. This is Mach's Principle. This means absolute motion is self-generating from the relative effects of matter, without having to arbitrarily introduce Newton's Absolute Space (immaterial universal field) or Maxwell's Luminiferous Aether (material field).

Einstein's theory of Special Relativity is blatantly inconsistent. Consider his example of polar and equatorial clocks on the earth. The equatorial clock is time dilated relative to the polar clock due to the earth's rotation. However, there is no reciprocal dilation effect *per-se* between the clocks as required by the theory – the effect is asymmetric and the polar clock runs fast relative to the equatorial clock. Also, Einstein's equations are covariant and there is nothing in his theory that mathematically derives the asymmetry, other than by the physicist's interference. And, the physical dilation of the equatorial clock contradicts Einstein's interpretation of the Michelson-Morley null result as being due to the earth observers being "at rest because there is no absolute motion". But if the equatorial clock is physically time dilated, then the Michelson-Morley apparatus cannot be "at rest" - it must be physically Lorentz contracted due to the earth's absolute motion, both rotational and orbital. The same reasoning applies to the Hafele-Keating Experiment, which also requires clocks on the earth to be time dilated, and not just aircraft clocks.

But what could this absolute motion be relative to? I say this is the average momentum of the Universe. The Michelson-Morley apparatus undergoes Lorentz contraction, which counters the "Aether Wind" effect. This keeps the speed of light constant for earth's observers, maintaining the Principle of Constant Light Speed.

Thus, it can be seen that Einstein's formulation of the Relativity Principle (First Postulate of Special Relativity) is false. This states the Laws of Physics must be "the same" for different inertial systems, and that this is so because his equations are covariant (symmetric). But that is false - the Laws of Physics can be the same even when there are asymmetries, and there is no restriction that different observers have to experience the same effects. This applies to both accelerated and inertial motion. Different examples can be given. For example, consider the annual stellar parallax seen by observers on earth. On other planets observers will measure a different change in parallax. We don't say "observers on different planets must experience the same parallax changes for the Laws of Physics to be the same". The Law of Physics is the same for all planets – the change in stellar parallax depends on a planet's orbital radius around the sun. So, observers on different planets see different parallax changes because the planets have different orbits. The same applies to Relativity. The Law of Physics for all clocks is this – time dilation is determined by a clock's absolute motion relative to the average momentum of all matter (the "Fixed Stars"). So, different clocks have different time dilations due to their different absolute motions. This means observers might measure a relatively moving clock to have a faster or slower rate (depending on their own absolute velocity), and relatively moving inertial observers can measure each other's clocks to have opposite rates. This is consistent with the rules of mathematics and logic, and applies to any form of motion. This is the Machian Relativity Principle.

The Principle of Machian Relativity shows that different inertial clocks do have different absolute speeds. This is an inevitable consequence of the combined influence of all matter. It must also be remembered there is no specific evidence for symmetric time dilation. All the experimental evidence is asymmetric, and no-one has proposed to specifically perform experiments on inertially moving bodies to see if the effect exists. So I propose three experiments to test the hypothesis. And from the evidence I give in this book, it is obvious that these experiments will give a null result for symmetry.

The status of Special Relativity being a special (inertial) case of General Relativity is also questionable. Special Relativity, in its original form, deals with straight line inertial motion, and is properly symmetric. Due to Einstein's Equivalence Principle, clocks free-falling in a gravitational field can be regarded as being inertial. Thus, Special Relativity is applicable, even though the motion is now curved. However, experiments with atomic clocks moving through the earth's gravitational field show that their motional time dilation is physically asymmetric.

Supposedly, there is no absolute motion because "no experiment conducted within an inertial system can detect it". For example, the Michelson-Morley experiment didn't show the earth's motion relative to the Aether, thus there was no Aether or absolute motion. However, the reality of the earth's absolute motion can be deduced by observing the external universe. We know the apparent retrograde motion of Mars arises from both the orbits of Mars and the Earth about the sun. Thus the earth cannot be regarded as being at rest, because its orbital motion is absolute. This can also be deduced from stellar parallax relative to the earth. According to Machian Relativity, the earth has absolute motion relative to the external universe (and we know from stellar parallax the earth has absolute motion relative to its light also). Therefore, Michelson-Morley's null result is due to the earth and the interferometer being physically Lorentz contracted by their absolute motion. De Sitter's Binary Star Principle tells us that light-spheres expand independently of their source motions. So, all light must have absolute motion relative to the "Fixed Stars". This must also mean that the light-sphere from the interferometer is not being emitted uniformly relative to the earth, thereby providing further proof of the earth's absolute motion. Thus, for observers to make deductions about their system, solely from observations within that system, is false. External observations and measurements must also be included. This applies not only to distant stars, but systems passing each other closely. [The absolute nature of the earth's orbital motion is no different from the absolute nature of its daily rotation. For example, rockets are launched in the direction of the earth's rotation to gain extra angular momentum.]

Absolute inertial motion must also occur in electromagnetism. Two inertially moving streams of parallel electric charges experience a magnetic attraction, which can overcome their mutual electric repulsion. Thus the currents eventually touch each other. This is an objective fact that applies to any observer (including one moving with the currents), and proves that inertial motion is absolute, but not relative as conventional science would claim.

Also, if conventional science claims that "acceleration causes an asymmetry", then at what point does the asymmetry vanish, if the acceleration of a moving clock is gradually reduced to zero? The equations must remain fully asymmetric, even if the acceleration becomes infinitesimally small (Fig.4):

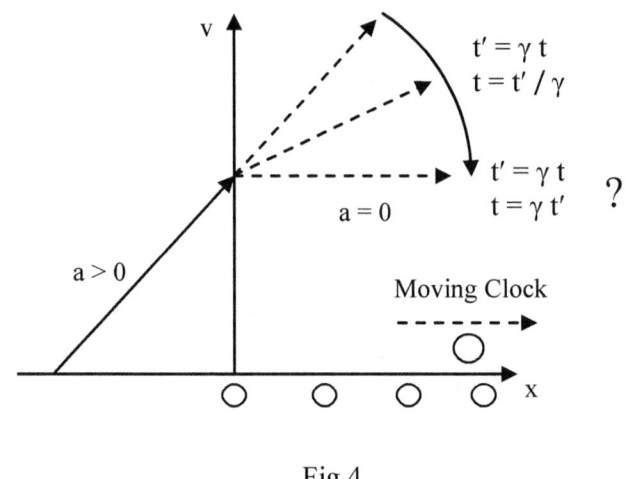

Fig.4

But according to Conventional Science, we must then switch from either General Relativity or "asymmetric Special Relativity", to symmetric Special Relativity, the moment the acceleration vanishes (a = 0). However, Machian Relativity tells us that the asymmetry ($t' = \gamma\, t$, $t = t' / \gamma$) persists, even when relative motion becomes inertial. This is the simplest outcome. So, space-time asymmetry persists regardless of a system's inertiality (inertial state).

And if different inertial frames measure the speed of light to be the same, it does not follow that their time dilations are reciprocal, or that light-spheres expand uniformly relative to their respective systems. Such ideas are false.

Glyn Phillips

Change in Relativity

Consider two identical balloons, A and B, inflated to the same size. Let some air be released from B. Relative to A, B is now smaller, but relative to B, A is bigger. So the changes are *asymmetric*. We don't say "because the balloons have changed relative to each other, then they must be smaller than each other, because everything is relative, and some covariant equation tells us so, and the Laws of Physics are the same". The change in B is absolute, because it has lost air. But the change in A is Relative or Apparent, because it has not lost air. So the cause of the change lies in B. The same applies to clocks. If clock B works *slower* than clock A, then A must work *faster* relative to B. I say, this reasoning applies if the clocks are stationary relative to each other (e.g. a conventional mechanical adjustment), and if they are in relative inertial motion (e.g. Time Dilation, a relativistic change). This is simpler than in conventional science, where inertial Time Dilation is somehow different, and has to be symmetric. Such inertial asymmetry is consistent with other clock rate asymmetries, such as during non-uniform motion, or clocks at different gravitational potentials.

Now consider a third identical balloon C, which loses the same amount of air as B. Relative to either balloon, there is no relative change in size. But this does not mean they have not changed, as both are smaller relative to A. The same applies to Time Dilation. All processes within a system are equally affected (e.g. quartz clocks, atomic clocks, heart rate, perception, etc). So, there is no relative change within the system. But this does not imply the system is somehow "at rest", as these processes can be compared with processes in a second system external to the first.

Now consider the balloons with the same size whose separation is then increased. Relative to balloon A's observer, B looks smaller, and relative to B's observer, A looks smaller. They only *look* smaller relative to each other because of perspective; neither has actually physically changed. So, the changes are *symmetric* and *apparent*. But it doesn't follow that because this example is symmetric, then other examples (e.g. air release) should also be symmetric.

11

Physical and apparent changes can also occur in combination. If air is released from B, and the separation increased, then relative to an observer with A, B is smaller due to a combination of physical decrease and perspective. The same applies in Relativity. Time dilation is physical and asymmetric, but can occur in combination with time delays, which are apparent.

The Scientific Literature

Scientists have never stated whether they think time dilation experiments have specifically demonstrated the reciprocal aspect of Special Relativity. This is a strange thing, when you consider the tremendous amount of speculation they have devoted to other areas, such as worm holes, dark energy, parallel universes, string theory, extra dimensions, the Higgs Boson, gravitational waves, miniature black holes, quantum foam, the arrow of time and whether the universe is a hologram or a computer simulation. The lack of clarity is also perplexing considering the billions of taxpayers' money spent by physicists globally on experimental research, such as particle physics, astronomy and space exploration. Such vagueness in any other field of knowledge would never be tolerated.

I have searched on the Internet for evidence of reciprocal dilation and found nothing. Wikipedia, which must be regarded as an authoritative source, only describes the (slower) rate of an aircraft clock relative to an earth clock in its article on the Hafele-Keating Experiment. Articles on time dilation experiments never describe what the corresponding rate of the earth clock relative to the aircraft clock might be, or mention words such as "covariant" and "reciprocal". The same applies to articles on GPS satellites. In order to "explain" the asymmetric (physical) dilation of each clock, both are classified as "non-inertial" because they are in rotating frames. The earth clock has a lower rotational velocity and physical dilation than the aircraft clock, thus it must run *fast* relative to the aircraft clock. If nothing is being said about evidence for reciprocal time dilation, and it has no practical applications, then this is telling us it DOES NOT EXIST.

BETA

EINSTEIN'S THEORY OF SPECIAL RELATIVITY

BETA I
The System

In relativity, observers in two relatively moving systems measure the speed of light from the same source and then compare their measurements. Each system consists of clocks and rigid rulers, which can be constructed in a variety of ways (e.g. this can be in the form of a square two–dimensional grid, corresponding to the x–y plane of a Cartesian coordinate system). The clocks are arranged in rows and columns, with rulers (rods) of the same length fixed between them. The clocks are synchronised with each other, and run at the same rate. Light sources and detectors are placed by each clock. Observers can also be positioned with each clock.

Inducement of Relative Motion:
Factoring out Acceleration

Consider two identical systems S_A and S_B of dimensions d, whose clocks have been mutually synchronised with those in the other system. Their origins are separated by a distance D, and their axes have the same orientations in space. A force is briefly applied to one of the systems, causing it to accelerate in the direction of the other system. The force is then released, and the system coasts inertially towards the other at constant speed v, according to Newton's First Law (Fig.5):

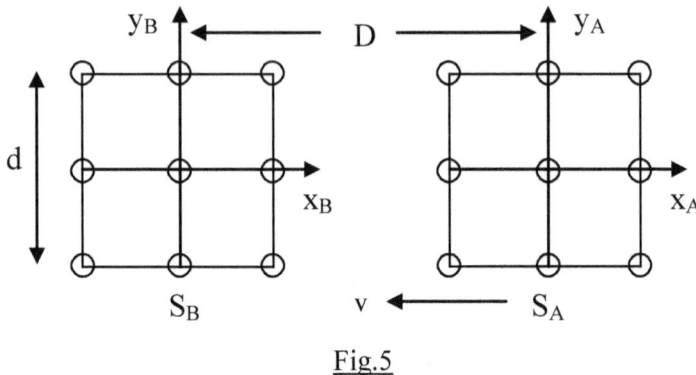

<u>Fig.5</u>

An important point of this procedure is that any acceleration <u>has stopped</u> before any comparison of measurements is made. This factors out accelerations, and ensures only inertial motion occurs. Both systems are therefore inertial, and in relative motion to each other. If we wish, we can further remove any doubt of asymmetry between both systems by subjecting the other system to an equal and opposite acceleration. Either way, forces are not acting during comparison of clock and ruler measurements, and forces and accelerations <u>are therefore irrelevant</u>. Because both systems are continuously moving, any changes in behaviour must be occurring during the motion itself, so must therefore be the result of that motion. I formulated this method to remove any ambiguity.

As a result, they inertially approach each other along the direction of their x–axes with the relative velocity v. The two systems pass each other in their immediate vicinity (they overlap). This reduces time delays between the two systems, allowing relatively moving clocks (and rulers) to be directly compared. Spherical light pulses (light–spheres) are emitted from the origins of both systems when they coincide. This corresponds to time t = 0 for both systems. Both systems continue moving inertially and recede from each other, each measuring the speed of light from both sources, and each directly comparing adjacent clock and ruler readings of the other system to their own.

Deriving the Equations

According to Einstein's 2nd Postulate, the speed of light c for both inertial systems must be constant regardless of the source motion. Assuming each system emits a spherical light pulse when their origins coincide, this means:

$$x_A^2 + y_A^2 + z_A^2 = c^2 t_A^2$$ (Expansion of both light–spheres relative to S_A)

$$x_B^2 + y_B^2 + z_B^2 = c^2 t_B^2$$ (Expansion of both light–spheres relative to S_B)

For simplicity only the x–axes are considered, as these are parallel with the relative velocity v between the systems. It is assumed that the transverse coordinates y and z remain unchanged for both systems. Thus the above equations simplify to:

$$x_A^2 - c^2 t_A^2 = x_B^2 - c^2 t_B^2 = 0 \qquad \{1\}$$

As there is one relative velocity for both systems, O_B (the origin of S_B) has the coordinates $x_A = v t_A$ in the system S_A, and O_A (the origin of S_A) has the coordinates $x_B = -v t_B$ in the system S_B. The simplest relation between the coordinates (x_A, t_A) and (x_B, t_B) can be written as:

$$\{2\} \qquad x_B = k_A (x_A - v t_A) \qquad \{3\} \qquad x_A = k_B (x_B + v t_B)$$

These equations $\{1, 2$ and $3\}$ can be solved to find k_A and k_B. The solution is described in textbooks on Special Relativity, and I will not repeat it, for two reasons. Firstly, this book is not intended to be a textbook on this theory. Secondly, I have checked the derivations myself and they are mathematically correct. My objective is to question Einstein's subsequent use of the equations in deriving the relativistic behaviour of clocks and rulers in relative motion (i.e. the physical interpretation of the equations).

The solution shows that $k_A = k_B = 1 / \sqrt{(1 - v^2/c^2)} = 1 / \gamma$. The previous equations then become:

{4a} $x_B = (x_A - vt_A) / \gamma$ {4b} $t_B = (t_A - x_A v/c^2) / \gamma$
{5a} $x_A = (x_B + vt_B) / \gamma$ {5b} $t_A = (t_B + x_B v/c^2) / \gamma$

The above equations are known as the Lorentz Transformations. These equations describe the measurements made by both systems as they measure the same thing – the speed of light from two simultaneous spherical pulses. Equations {4a, 4b} are the measurements (x_B, t_B) made in S_B due to its motion relative to S_A. Equations {5a, 5b} are the measurements (x_A, t_A) made in S_A due to its motion relative to S_B. In contrast, Galilean Relativity has:

$$x_B = (x_A - vt_A) \qquad\qquad t_B = t_A$$
$$x_A = (x_B + vt_B) \qquad\qquad t_A = t_B$$

The reason why $x_B \neq (x_A - vt_A)$, $x_A \neq (x_B + vt_B)$ and $t_A \neq t_B$ for Special Relativity is because each system undergoes physical changes due to its relative motion (unlike in Galilean Relativity). This is the reason why they measure c to be constant, but not c + v, or c − v (as for classical relativity). So space and time are no longer absolute (constant), as in Galilean Relativity. However, for low relative velocities v / c ≈ 0, so $\gamma = 1$ and the above equations approximate to those of classical (Galilean) relativity. Also as v > c, $\gamma > 0$ and $1 / \gamma > \infty$.

The Lorentz-FitzGerald Length Contraction

It can be shown that the lengths of objects in each system contract in their direction of motion as they move relative to the other system. Let the system S_A measure the length of an object. Let the position of one end be $x_A 1$, and let the position of the other end be $x_A 2$. For $x_A 1$, let the corresponding measurement made by S_B be $x_B 1$, and for $x_A 2$ let this be $x_B 2$. Using the Lorentz Transformation equation we have:

$$x_B 1 = (x_A 1 - vt_A 1) / \sqrt{(1 - v^2/c^2)}$$
$$x_B 2 = (x_A 2 - vt_A 2) / \sqrt{(1 - v^2/c^2)}$$

In S_A, the measurements are made simultaneously, so that $t_A 1 = t_A 2$. The length measured by S_A is $X_A = x_A 2 - x_A 1$, and the corresponding length measured by S_B is:

$$X_B = x_B2 - x_B1 = (x_A2 - x_A1) / \sqrt{(1 - v^2/c^2)}$$
$$= X_A / \sqrt{(1 - v^2/c^2)}$$

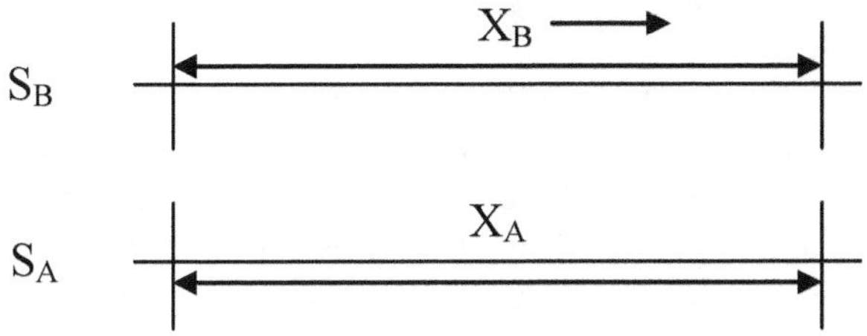

$$S_B \qquad X_B \longrightarrow$$

$$S_A \qquad X_A$$

Fig.6

In other words, when S_A and S_B measure the length of the same thing (Fig.6), the measurement made by observers in S_B is greater than the corresponding measurement by S_A's observers. That is, the measured lengths are not equal as predicted by Newtonian–Galilean relativity. The difference in S_B's measurement must arise from a physical contraction in its rulers as it moves relative to S_A – because S_B's rulers are now smaller, they give a higher reading when both systems measure the same thing. From the previous equation for measurements, the length contraction of S_B's rulers is therefore:

$$L_B = L_A \sqrt{(1 - v^2/c^2)}$$

It can be seen that when $v = 0$, $L_B = L_A$, and that the contraction of L_B increases with v, until $L_B = 0$ when $v = c$. This affects all objects in L_B, so that observers in L_B cannot detect this change by measuring their own objects, but this does not mean that nothing has happened to S_B – if nothing had happened, its measurements would be determined by classical relativity and its measured speed of light would not remain invariant. These equations are derived from S_B's measurements, and must represent an objectively real physical change in S_B – to say that they are only "relative to S_A" is completely false – the change is not just a subjective change in S_B's appearance as viewed by S_A's observers.

The change of lengths in S_A due to its motion relative to S_B is solved in exactly the same way from the Lorentz Equations:

$$x_A1 = (x_B1 - vt_B1) / \sqrt{(1 - v^2/c^2)}$$
$$x_A2 = (x_B2 - vt_B2) / \sqrt{(1 - v^2/c^2)}$$

Einstein assumes that S_B is exactly equivalent to S_A because they are both inertial, so that observers in S_B consider themselves to be at rest. This is Einstein's 1st Postulate of Special Relativity (the Principle of Relativity). Therefore, they measure the positions of each end of an object simultaneously; $vt_B1 = vt_B2$. This gives:

$$X_A = X_B / \sqrt{(1 - v^2/c^2)}$$
And $\quad L_A = L_B \sqrt{(1 - v^2/c^2)}$

So we can see that not only do objects in S_B contract due to motion relative to S_A (Fig.7a), but objects in S_A also contract due to motion relative to S_B (Fig.7b):

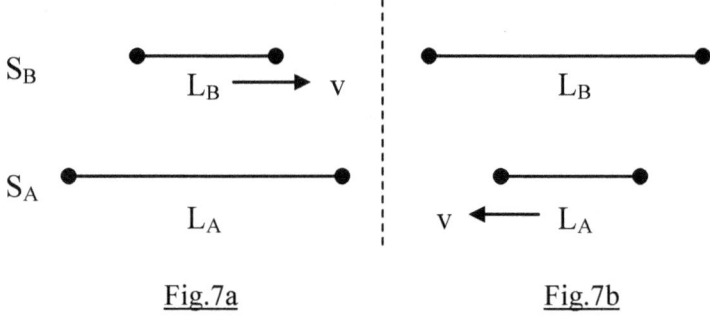

Fig.7a   Fig.7b

So there is complete symmetry between S_A and S_B – not only the Lorentz Equations, but the resulting Lorentz Contractions, are covariant. Physicists say; "the Laws of Physics are the same for different inertial frames".

Time Dilation

From the Lorentz Equations, it can be shown that the passage of time in each system is slowed as it moves relative to the other system. When the origins of both systems coincide, $x_A1 = x_B1 = 0$, and the clocks at the origin of each system read $t_A1 = t_B1 = 0$. For S_B moving relative to S_A, we have from the Lorentz Equation {4b}:

$$t_B1 = (t_A1 - x_A1v/c^2) / \sqrt{(1 - v^2/c^2)}$$
$$t_B2 = (t_A2 - x_A2v/c^2) / \sqrt{(1 - v^2/c^2)}$$

Subtracting:

$$(t_B2 - t_B1) = (t_A2 - t_A1 - x_A2v/c^2 + x_A1v/c^2) / \sqrt{(1 - v^2/c^2)}$$

The duration for S_A is $\Delta_A = (t_A2 - t_A1)$ and for S_B is $\Delta_B = (t_B2 - t_B1)$. Relative to S_A, S_B's origin travels a distance $(x_A2 - x_A1) = v(t_A2 - t_A1) = x_A2 = vt_A2$. Thus:

$$\Delta_B = (\Delta_A - t_A2v^2/c^2 + t_A1v^2/c^2) / \sqrt{(1 - v^2/c^2)}$$
$$= (\Delta_A - \Delta_Av^2/c^2) / \sqrt{(1 - v^2/c^2)}$$
$$= \Delta_A \sqrt{(1 - v^2/c^2)}$$

Thus, the duration measured by a clock in S_B as it moves relative to S_A is less than the corresponding duration measured by clocks in S_A (Fig.8):

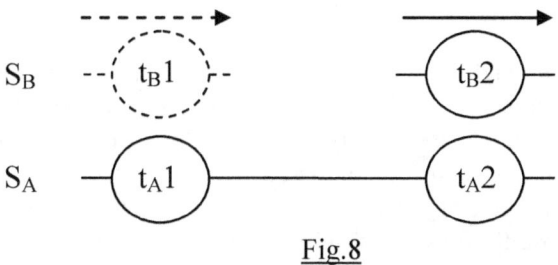

Fig.8

If the duration measured by S_B is less than the corresponding duration measured by S_A, this means that clocks in S_B are running at a slower rate than clocks in S_A. In other words, from the difference in measurements, we can deduce that a physical change has occurred. From the equations for durations, the rate of S_B's clocks relative to S_A's clocks is:

$$T_B = T_A \sqrt{(1 - v^2/c^2)}$$

The rate of clocks in S_A due to motion relative to S_B is solved in exactly the same way, because (according to Einstein) S_B is equally at rest as S_A is. The solution is identical to the previous one, and predicts S_A runs at a slower rate due to motion relative to S_B. Therefore we have:

$$T_B = T_A \sqrt{(1 - v^2/c^2)} \qquad T_A = T_B \sqrt{(1 - v^2/c^2)}$$

That is, the equations for time dilation are also covariant – clocks in S_B run slower due to motion relative to S_A, and vice versa. The rate of a clock in S_B is deduced by comparing its readings to clocks in S_A as it passes them. Exactly the same thing is done to deduce the rate of a clock in S_A due to its motion relative to S_B:

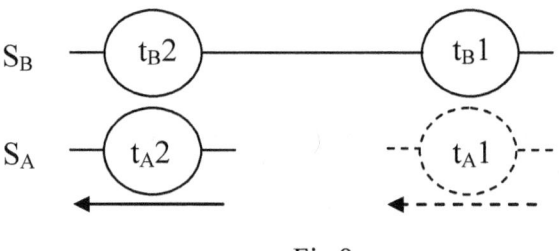

<u>Fig.9</u>

Sets of clocks in each system are assumed to remain synchronized relative to their respective observers, because they are both inertial and (according to Einstein) "equivalent". Thus, a clock in one inertially moving system is compared to a set of clocks in the other, and so the "return journey" and "accelerations" of the Twins Paradox (supposedly for clocks to be compared) are irrelevant.

Relative Simultaneity

It can be shown that the simultaneity of clock readings in each system is different as it moves relative to the other system. If the origins of both systems are defined to be $t_A 1 = t_B 1 = 0$ when they coincide, the corresponding clock readings along the x–axis of each system are given by:

$$t_B2 = -vx_A / c^2\sqrt{(1 - v^2/c^2)} \qquad t_B2 = 0$$
$$t_A2 = 0 \qquad\qquad\qquad\qquad t_A2 = vx_B / c^2\sqrt{(1 - v^2/c^2)}$$
$$(S_B \text{ moving rel to } S_A) \qquad\qquad (S_A \text{ moving rel to } S_B)$$

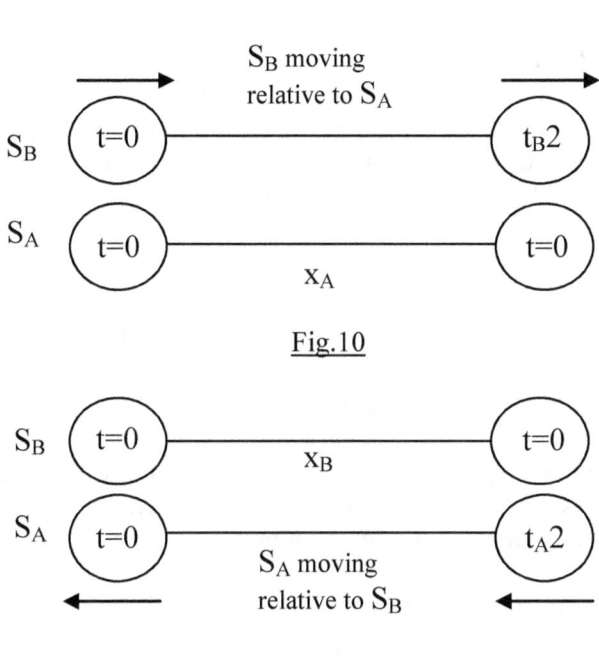

Fig.10

Fig.11

The synchronization difference in clock readings within each moving system is maintained relative to the origin of that system, and therefore moves with the system. This ensures that each (moving) clock maintains its dilated clock rate relative to the other (stationary) system according to the previous time dilation equations.

The notion of time delays (frequently mentioned in other books) in the derivation of Relativistic Simultaneity, Lorentz Contraction and Time Dilation is irrelevant. This is because clocks and rulers in both systems are directly compared to each other in their immediate vicinity as they pass each other.

A brief description of other relativistic phenomena will be given. A more detailed analysis can be found in the numerous books on relativity published since Special Relativity was first developed.

Relativistic Mass Increase

In Special Relativity, the masses of bodies in a system change due to motion relative to another system. Consider systems S_A and S_B as before, initially at rest with each other. In S_A let there be a mass m_A, and in S_B a mass m_B, such that $m_A = m_B$. If S_B moves inertially (or non-inertially) relative to S_A, m_B varies relative to m_A according to:

$$m_B = m_A / \sqrt{(1 - v^2/c^2)}$$
Or: $\quad m = m_0 / \sqrt{(1 - v^2/c^2)}$ (standard notation)

Thus as a system's relative velocity approaches the speed of light, its mass approaches infinite magnitude. This makes a mass harder to accelerate, and so no system can reach or exceed the speed of light. In a system, the energy of a body is proportional to its mass, according to $E = mc^2$. Thus the previous equation becomes:

$$E_B = E_A / \sqrt{(1 - v^2/c^2)}$$
Or: $\quad E = E_0 / \sqrt{(1 - v^2/c^2)}$ (standard notation)

Also because of the Relativity Principle, S_A and S_B must be equivalent when S_B moves inertially. Thus mass and energy in S_A must change reciprocally relative to S_B according to:

$$m_A = m_B / \sqrt{(1 - v^2/c^2)} \qquad E_A = E_B / \sqrt{(1 - v^2/c^2)}$$

Relativistic Velocity Addition

Special Relativity deals with various scenarios regarding how processes measured in one inertial system are measured by another relatively moving inertial system. This includes an inertial body or system moving perpendicular to the relative motion of the two systems, or a body or system undergoing acceleration relative to the two inertial systems. The simplest scenario is when a third system moves inertially relative to the first two, but in the same direction. For example, if relative to the x–axis of S_A, S_B has velocity v and S_C has velocity –u, the combined relativity velocity of S_C relative to S_B is not w = v + u according to Galilean Relativity, but is given by:

$$w = (v + u) / (1 + vu/c^2)$$

In addition to the previous result (where a system moving relative to another cannot reach or exceed the speed of light), this prevents a combined relative velocity between two systems reaching or exceeding the speed light. For example, if v = u = 0.9c, then we have w = 1.8c/ (1 + 0.81) < c.

Corollary – On the Physicality of the Equations

The change in a system *is deduced from the measurements made by observers in the system itself,* so the change must be <u>physical and objectively true for both systems</u>. We cannot say the changes in a system are only relevant for observers in another system, and thereby evade any paradoxes by implying the changes are only a form of subjective illusion or process for such observers.

So, these equations describe proper physical changes. The changes are not some form of apparent or optical effect that is only true for an observer, and the variables (x_A, t_A) and (x_B, t_B) represent the respective measurements made by each system of a light pulse.

Further proof of the physicality of the theory can be demonstrated by considering light clocks. Let systems S_A and S_B consist of light clocks, moving relative to each other as described previously. When the origins O_A and O_B pass in each other's immediate vicinity, they emit simultaneous light pulses. Let system S_A measure light from its own source to be a sphere centred on O_A expanding outward with speed c. It can be shown using Pythagoras's Theorem that if the sphere is of a certain size at time t_A in system S_A, then the corresponding reading on S_B's clock is t_B = t_A $\sqrt{(1 - v^2/c^2)}$ which results in S_B measuring the same light pulse to have speed c also. The reading t_B is an objective change that is true for observers in both S_A and S_B. Likewise; the reading

t_A on clocks in S_A is objectively true for observers in S_A and S_B. Therefore, observers in S_A conclude that clocks in S_B are time dilated (running slower), while observers in S_B conclude that clocks in S_A are anti–dilated (running faster). This faster rate is consistent with the result for atomic clocks on airliners, when the clock is flown east–west, while the slower rate is consistent with clocks being flown west–east.

Corollary

Acceleration (mentioned in other books discussing the Twins Paradox) is irrelevant to the theory because the systems move inertially throughout their comparisons. Einstein's original method compares a clock in one system which instantly starts moving from a clock in the other system and then instantly stops at another. Because the accelerations are instantaneous, they have no duration and therefore no other effect on the process. However, the stopping and starting might make the presence of acceleration in the theory ambiguous, so I have completely factored it out it in my derivation. The systems are continuously moving throughout the comparison process, so the changes must be occurring during inertial motion.

Corollary – The meaning of "v"

The symbol "v" in these equations is the relative velocity, that is, the velocity of S_A relative to S_B and of S_B relative to S_A. If the velocity of S_B relative to S_A is v_A and the velocity of S_A relative to S_B is v_B, then $v = v_A = v_B = v_R$.

In Lorentz's Aether relativity, the symbol "v" refers to velocity of a system relative to the Aether, not between systems. In this case, we have $v = v_E$.

In Machian Relativity, there is no Aether. Instead, the relativistic effects of all bodies in the universe produce an average effect in proportion to their respective masses. In this case, velocity is relative to the Plenum, so $v = v_P$. This is approximately the velocity relative to the earth, as celestial bodies such as stars and planets have low absolute velocities due to their large masses.

Corollary

In deriving the above equations, it is assumed that each inertial system can consider itself at rest – this is the nature of Einstein's Relativity Principle (Postulate of Inertial Equivalence). Accordingly, both sets of equations for systems S_A and S_B are derived in the same way. This generates the symmetry in the above equations, and this is known as covariance. However, I say that the use of symmetry to make the Laws of Physics "the same for both systems" is bogus. The Laws of Physics are still the same even for asymmetric phenomena. Consider for example, two cyclists – one stays beside the road and experiences no wind, while the other moves at constant speed and experiences a wind. The Law of Physics is the same for both – the air speed is proportional to the cyclist's speed relative to it. So the first cyclist experiences no wind speed because he is not moving relative to the air. We don't say, "Both cyclists are moving inertially relative to each other, so they must experience the same wind speed for the Laws of Physics to be the same".

Also, the symmetry of Einstein's equations for both inertial systems violates the rules of algebra. Acceleration cannot be used as an excuse because it has been factored out. And claiming the equations are "unphysical" is fallacious. Also, if X changes relative to Y, claiming "X's change is only true for observers in Y but not those in X" is rhetorical nonsense that confuses physical changes with apparent changes (like parallax).

Asymmetry is a better way of making the Laws of Physics the same for both systems, as the equations obey the rules of algebra. According to Machian Relativity, asymmetry is relative to the average momentum of all matter in the universe. This means that observers in one inertial system can detect their absolute motion, by observing phenomena external to their reference frame, and so conclude that they are not at rest. In other books you might read that "no experiments in an inertial system can detect absolute motion". But it is erroneous to restrict observations in this way. For example, the equations of Special Relativity are derived by direct comparison of two different systems. On the contrary, I say that no experiment will ever detect the symmetric time dilations predicted by Einstein's theory, but only the asymmetric dilations of Machian Relativity.

BETA II
Accelerations and Asymmetry in Special Relativity

Asymmetry in the equations during acceleration is never derived from first principles, but is only ever assumed. For the Twins Paradox (discontinuous acceleration), the outward and return journeys are mostly inertial, and the corresponding clock rates (and twins' ageing) are assumed to be symmetric according to Special Relativity. The resulting asymmetry of the travelling twin on his return is explained in a variety of ways depending on which book you read. Some say General Relativity is the answer, while others say General Relativity is not needed, and instead combine Special Relativity with the idea of Time Delays (contradicting Einstein's original method of using systems of rulers and synchronised clocks to factor out such effects). For <u>continuous acceleration</u> (particles orbiting in accelerators), the equations of Special Relativity are used in an asymmetric way without any mention of Time Delays or General Relativity, thereby forcing the equations to match the experimental results, which are asymmetric. This is said to "experimentally prove" Special Relativity, but the hypothesis of symmetry during inertial motion has not been tested. In addition, relativistic mass increase in Special Relativity prevents bodies from being accelerated to the speed of light. So in this case at least, Special Relativity appears to apply to acceleration.

Also, during their acceleration, non–inertial observers should measure anti–dilations and anti–contractions in the inertial system (due to the asymmetry of the physical change). Therefore, there should be corresponding equations describing such changes, but they are never mentioned by conventional science.

However, let us assume that the hypothesis of acceleration asymmetry is true. If the system S_B is continuously accelerated by applied forces relative to S_A (which remains inertial), then the previous equations become:

$$L_B = L_A \sqrt{(1 - v^2/c^2)} \qquad L_A = L_B / \sqrt{(1 - v^2/c^2)}$$

$$T_B = T_A \sqrt{(1 - v^2/c^2)} \qquad T_A = T_B / \sqrt{(1 - v^2/c^2)}$$

Unlike the previous covariant equations derived for inertial motion, these new asymmetric equations <u>obey the laws of algebra and logic</u>. However, it must be remembered that they have been assumed, not proved by derivation. The symbol "v" now represents the instantaneous velocity, and T and L represent instantaneous durations and lengths.

Also, if the system S_B is given a constant acceleration over a particular duration, then for low values of v the above equations can be written as:

$$\langle L_B \rangle = L_A \sqrt{(1 - \langle v \rangle^2/c^2)} \qquad \langle L_A \rangle = L_B / \sqrt{(1 - \langle v \rangle^2/c^2)}$$

$$\langle T_B \rangle = T_A \sqrt{(1 - \langle v \rangle^2/c^2)} \qquad \langle T_A \rangle = T_B / \sqrt{(1 - \langle v \rangle^2/c^2)}$$

Where: $\langle L \rangle$, $\langle T \rangle$ and $\langle v \rangle$ are the respective average values of the previous variables over the duration.

In addition, if we say "accelerations cause asymmetries" then the equations are asymmetric for any magnitude of acceleration, even very small ones. Therefore, if the acceleration of system S_B is made infinitesimally small, the asymmetry of the equations must remain. Therefore, system S_B can be considered to have a constant speed v throughout the process, and yet the equations must remain asymmetric:

$$L_B \approx L_A \sqrt{(1 - v^2/c^2)} \qquad L_A \approx L_B / \sqrt{(1 - v^2/c^2)}$$

$$T_B \approx T_A \sqrt{(1 - v^2/c^2)} \qquad T_A \approx T_B / \sqrt{(1 - v^2/c^2)}$$

Furthermore, the asymmetry is reversed if instead system S_A is accelerated by applied forces while S_B remains inertial. In this case, system S_B now defines the state of rest. The problem with this method is its arbitrariness. Firstly, the asymmetry is assumed by ignoring one system's set of symmetric equations because it is non–inertial. Secondly, the asymmetry is assumed to be relative to the inertial system it was originally at rest with, but this is equally arbitrary. If a system is accelerating relative to the earth, it is also accelerating relative to the sun, so there is no clear choice of rest system to set the asymmetry to. Thirdly, there is also the issue of what happens to the asymmetry when the acceleration vanishes, and S_B moves inertially.

However, the Theory of Machian Relativity allows the asymmetry to be derived from First Principles, directly from the Lorentz Equations. It also clearly defines what the asymmetry is relative to – the average momentum of all matter, which defines a state of absolute rest. In addition, I say that the asymmetry remains when the acceleration vanishes (inertial motion). This is a simpler outcome than conventional science.

BETA III
The Wavefront Paradox of Special Relativity

A further ambiguity in Einstein's method is how light diverges from each source. The conventional method says "a pulse of light is emitted when the origin of both systems coincide", without considering from which source the light is coming from. For example, if light is emitted from S_A's source, we assume the spherical wavefront Ψ_A remains centred on the origin of S_A. But what happens to the corresponding pulse from the source in S_B? Which system is the wavefront Ψ_B centred on?

In the diagram for Special Relativity (Fig.12), it can be seen that there is a paradox – both wavefronts are centred on S_A because it is "at rest", in which case the space-time of S_B changes to maintain c-invariance for its observers. But then S_B is also "at rest", because "inertial frames are equivalent", which means that both wavefronts must be centred on it, so that S_A's space-time changes instead (contradicting the previous conclusion). Consider also the Light Clock argument used in explaining time dilation. The predicted time dilation for clocks in one system is based on the assumption that both light spheres remain centered on the other system.

a) Special Relativity: Wavefronts paradoxically centred on each system because they are "equivalent":

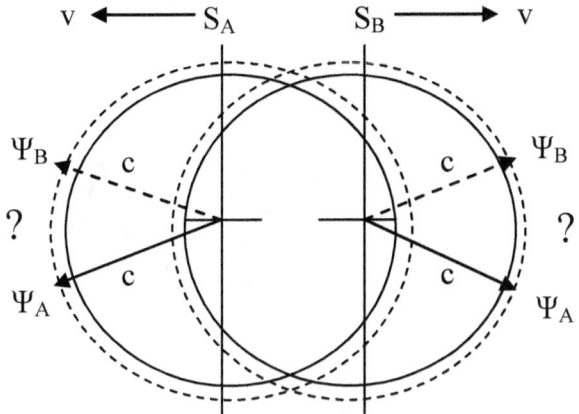

Fig.12: Paradoxical Wavefronts: Special Relativity

This is another manifestation of the paradoxes that arise in Special Relativity, which have never been adequately resolved. Other relativity theories avoid this paradox in different ways, although they do not necessarily preserve c-invariance:

b) Wavefronts Ψ_A and Ψ_B centred on S_A and S_B respectively (c is constant relative to its source only):

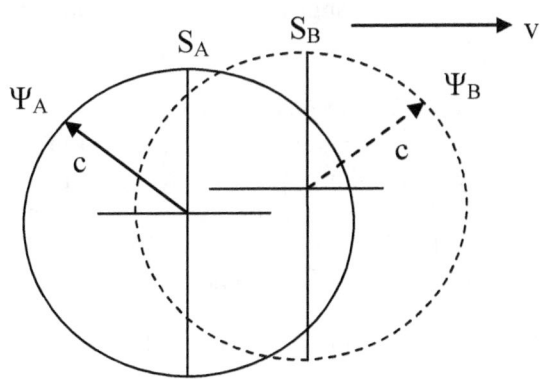

Fig.13: Paradox-free Wavefronts: Newtonian–Galilean Relativity

c) Wavefronts Ψ_A and Ψ_B centred on S_A if at rest in the Luminiferous Aether Λ (c-invariance if S_B's space-time changes):

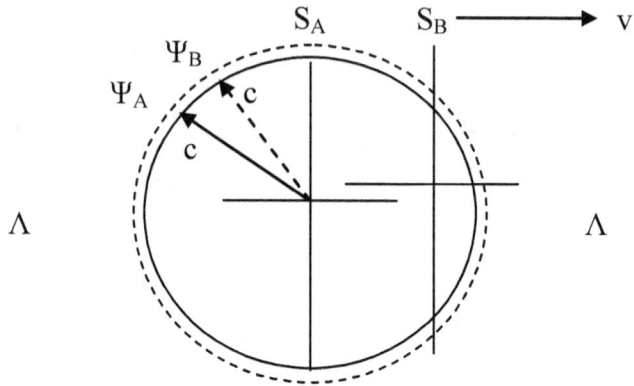

Fig.14: Paradox-free Wavefronts: Aether Relativity

Resolution of the Paradox:
The Postulate of the Mach Field

Consider the motion of S_B relative to S_A. The changes in S_B are a result of this motion. So, the time dilation of its clocks must persist in the space between S_A's clocks. This can be verified by placing additional clocks between S_A's existing clocks. The same reasoning applies to changes in S_A's clocks and rulers due to its motion relative to S_B.

If the two systems S_A and S_B continue in their inertial motion, they diverge from each other, getting further apart until no part of each system overlaps with the other. However, the relativistic changes (time dilations) for each system must persist, because the relative motions are continuing. This can be verified by placing additional clocks and rulers in the space around each system.

These facts have not found any place in the foundations of our edifice of the physical universe. But we arrive at a very satisfactory interpretation, if we assume that there is an immaterial causal field associated with each system, through which the other system moves and is thereby changed. I call this a <u>Relative Space</u> or <u>Mach Field</u>. Unlike a force field, a Mach Field is completely uniform throughout space without any spatial reduction in its "strength", and the magnitude of its effect (on the other system's space-time) depends on the speed of the other system relative to it (through its Mach Field). I also say the source of a system's Mach Field is its mass, so that the "strength" of its effect also depends on its mass. The concept of the Mach Field is inherent to Relativity, but has never been previously identified. Mach Fields are the causal mechanism for relativistic effects, and I also say that they combine to produce average effects.

It can be seen in the diagram for Machian Relativity that the Wavefront Paradox of Special Relativity is resolved (Fig.15).

d) Machian Relativity: Wavefronts Ψ_A and Ψ_B centred on centre of gravity (Γ_{AB}) of S_A and S_B:

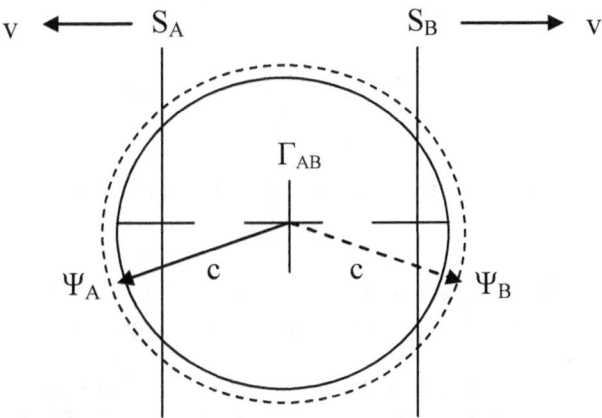

Fig.15: Paradox-free Wavefronts (two-body universe with equal masses)

Now, the mass of each system has an effect (by virtue of its Mach Field), with one system dragging both wavefronts one way, while the mass of the second system drags them in the opposite direction. By "dragging" I mean the tendency of a system's mass to keep light–spheres from both systems centred on itself – that is, to maintain absolute motion for light. This is not Stokes's Aether Drag Theory[1].

If both masses are equal, they have *equal and opposite* dragging effects, so that both wavefronts are centred on the midpoint of both systems. If the mass of one system is greater than the other, it has a *greater* dragging effect on the wavefronts. This means that the wavefronts are always centred on the *average momentum* of both systems (Fig.16):

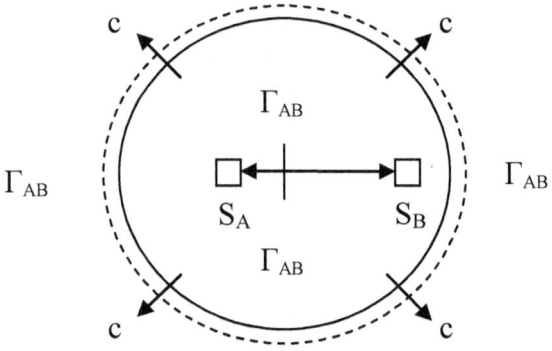

Fig.16: Paradox-free Wavefronts (two-body universe with unequal masses)

In other words, the absolute motion of light from both sources is relative to the average momentum, while the motion of light

---

[1] Note: Stokes's Aether Drag Theory requires complete dragging of light in the vicinity of the earth or other massive body. Such complete dragging gives a null result for the Michelson-Morley Experiment (e.g. c-invariance in the frame of the earth), without the additional requirement for Lorentz contraction or time dilation. Machian relativity produces a *universal average dragging effect* on light throughout *all space* that corresponds to a state of absolute rest. The earth has absolute motion relative to this state of rest, and experiences Lorentz contraction, which counters the "Aether wind effect", giving the required null result for the Michelson-Morley Experiment. So, Stokes's Aether drag is a *localised* effect only, while Machian dragging is *global. This theory is not compatible with Stellar Aberration so must be false.*

(expansion of light–spheres) relative to both sources is non–absolute (not centred on either system). This is The Machian Wavefront Theorem.

Each system then undergoes relativistic changes in its space–time, determined by its speed relative to the average momentum (and the centre of the lightsphere) – this means the speed of light relative to both systems is absolute (constant magnitude), even though its motion relative to both is non–absolute. From this, the mass effects of other objects also need to be included, and this leads to Mach's Principle – relativistic changes and the motion of light are determined according to motion relative to the *average momentum of all matter in the Universe* – the Cosmic Plenum of all Mach Fields, $\Gamma\infty$ (Fig.17):

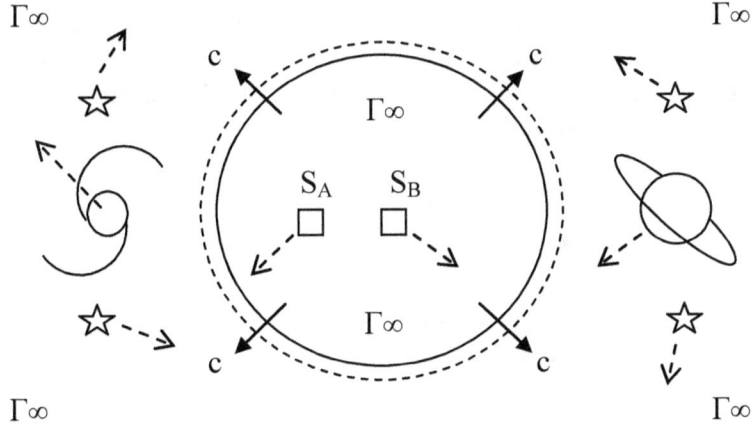

Fig.17: Paradox-free Wavefronts (average momentum of Universe)

According to Machian Relativity, light exists as photons in *a Plenum of immaterial Machian Fields of all matter*. There can be no epistemological objection to this, as we also know that space is a plenum of the immaterial Gravitational fields of all matter.

GAMMA

THE EQUATIONS OF MACHIAN RELATIVITY DERIVED BY ALGEBRAIC METHODS

GAMMA I
The Mach–Lorentz Transformations for a Two–body Universe

Consider the two identical systems S_A and S_B moving as described in the previous examples. Let this scenario represent a simplified two–body universe. This influence of other bodies (i.e. the external universe) will be considered later. As before, both systems emit spherical light pulses when their origins pass each other. Each system has the same mass and respective Mach Fields, Φ_A and Φ_B, each of which uniformly fills the space around both systems.

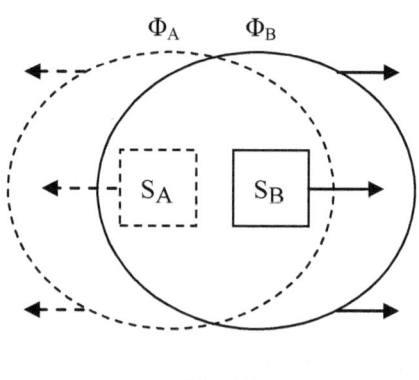

Fig.18

These fields mutually interact, dragging each lightsphere equally in either direction. Thus, light–spheres Ψ_A and Ψ_B (emitted from sources Σ_A and Σ_B at the origins of S_A and S_B respectively), expand uniformly (at speed c) relative to the average momentum of both systems (Machian Wavefront Theorem), and thus remain centred on the point Ω where both origins previously passed each other:

34

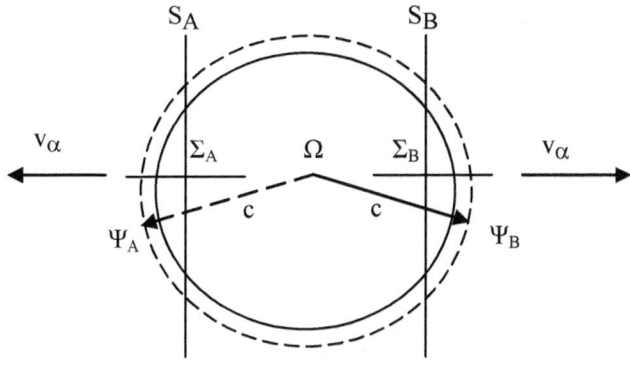

Fig.19

On the basis of these wavefronts, the space-time changes in both systems are calculated to keep the speed of light constant for observers in either system. Thus:

$$x_A^2 - c^2 t_A^2 = x_B^2 - c^2 t_B^2 = 0$$

So, instead of clocks and rulers in each system reciprocally changing relative to each other, <u>the two systems change according to their speeds relative to their common centre of gravity</u>. That is, they now change not due to their relative velocity v, but their velocity relative to the combined system [S_A+S_B], which is v/2. Let v/2 = v_α.

[Clocks and rulers in a system tend to change due to their speed v relative to another system of the same mass. However, they also tend not to change (self-influence) due to being at rest relative to their own system mass (v = 0). Thus, the average change is determined by v/2. QED: this outcome is identical to that of the previous argument based on wavefronts].

A third system of reference S_α is placed at the centre of gravity of S_A and S_B (Fig.20). Because the speed of light must be a constant for all inertial systems, S_α must measure the wavefronts emitted by S_A and S_B to have the same speed c:

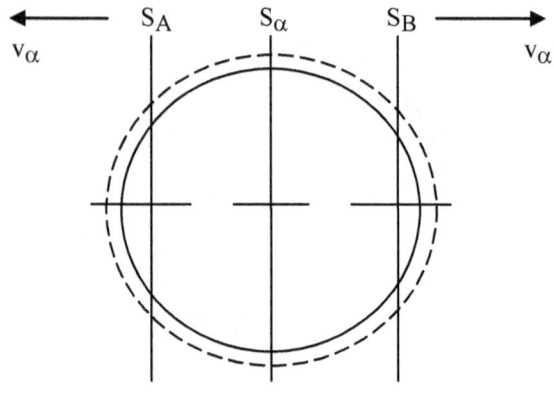

$$S_A \qquad S_\alpha \qquad S_B$$

$$v_\alpha \qquad\qquad\qquad\qquad\qquad v_\alpha$$

Fig.20

In fact, because S_α is at the centre of gravity, light from S_A and S_B has absolute motion relative to it and the *space-time of S_α is unchanged*. Therefore the above equation can now be written as:

$$x_A^2 - c^2 t_A^2 = x_B^2 - c^2 t_B^2 = x_\alpha^2 - c^2 t_\alpha^2 = 0$$

Initially consider the systems S_α and S_B. The simplest relation between the coordinates (x_α, t_α) and (x_B, t_B) can be written as:

$$x_B = k_\alpha (x_\alpha - v_\alpha t_\alpha) \qquad\qquad x_\alpha = k_B (x_B + v_B t_B)$$

As a further simplification, $v_B = v_\alpha$ (this will be proved later). Thus:

$$x_B = k_\alpha (x_\alpha - v_\alpha t_\alpha) \qquad\qquad x_\alpha = k_B (x_B + v_\alpha t_B)$$

These equations are solved in the same way as for Special Relativity. The solutions shows that $k_\alpha = k_B = 1 / \sqrt{(1 - v_\alpha^2/c^2)} = 1/\gamma$. So γ is now a function of absolute velocity v_α instead of relative velocity v. The previous equations then become:

$$x_B = (x_\alpha - v_\alpha t_\alpha) / \gamma \qquad t_B = (t_\alpha - x_\alpha v_\alpha/c^2) / \gamma$$
$$x_\alpha = (x_B + v_\alpha t_B) / \gamma \qquad t_\alpha = (t_B + x_B v_\alpha/c^2) / \gamma$$

These equations are the Lorentz Transformations for Machian Relativity. The corresponding transform equations for S_A are solved in exactly the same way as for S_B, and are identical to those above. So far, these equations have the same form as those for Special Relativity, but the physical processes behind them are different, and so is the method of their solution. That is, in Special Relativity observers in each frame (S_A and S_B) are "at rest", with each frame undergoing reciprocal changes relative to the other. However, in Machian Relativity, both frames S_A and S_B interact via their Mach fields, so that they change due to motion relative to their centre of gravity.

Machian Time Dilation

Let observers in both systems compare the readings of a clock in S_B when it consecutively passes two synchronised clocks in S_α located at $x_\alpha 1$ and $x_\alpha 2$. For $x_\alpha 1$, let the reading of S_B's clock be $t_B 1$ and the reading of S_α's clock be $t_\alpha 1$. For $x_\alpha 2$ these readings are then $t_B 2$ and $t_\alpha 2$.

It must be remembered that these readings are physically real and therefore objectively true for both systems. The readings on the clock in S_B are true for both S_α and S_B – they are not some form of subjective process that is only true for S_α. Using the Lorentz equation for t_B we have:

$$T_B = t_B 2 - t_B 1 = (t_\alpha 2 - x_\alpha 2 v_\alpha / c^2) / \gamma - (t_\alpha 1 - x_\alpha 1 v_\alpha / c^2) / \gamma$$

$$= (t_\alpha 2 - t_\alpha 1) / \gamma - v_\alpha (x_\alpha 2 - x_\alpha 1) / c^2 \gamma$$

The motion of the origin O_B relative to S_α is described by:

$$x_\alpha 1 = v_\alpha t_\alpha 1 \qquad x_\alpha 2 = v_\alpha t_\alpha 2$$

[That is, the clock that moves with O_B is at $x_\alpha 1$ at $t_\alpha 1$ and then at $x_\alpha 2$ at $t_\alpha 2$]. Therefore:

$$x_\alpha 2 - x_\alpha 1 = v_\alpha (t_\alpha 2 - t_\alpha 1)$$

By substituting these values into the previous equation for T_B, it can be shown that the measured duration (and rate) is:

$$T_B = \gamma T_\alpha$$

This is the same result as derived by the geometric method of Machian Relativity (p.51) and by Special Relativity.

Machian Lorentz Contraction

Let the system S_α measure a length X_α on its x–axis to be $x_\alpha 2 - x_\alpha 1$. Using the Lorentz Transformation for x_B, we have:

$$X_B = x_B 2 - x_B 1 = (x_\alpha 2 - v_\alpha t_\alpha 2) / \gamma - (x_\alpha 1 - v_\alpha t_\alpha 1) / \gamma$$
$$= (x_\alpha 2 - x_\alpha 1) / \gamma - (v_\alpha t_\alpha 2 - v_\alpha t_\alpha 1) / \gamma$$

In S_α the readings $t_\alpha 2 = t_\alpha 1$ because the position of either end is measured simultaneously. Therefore:

$$X_B = (x_\alpha 2 - x_\alpha 1) / \gamma = X_\alpha / \gamma$$

That is, the corresponding measurement of the same dimension by S_B has a greater value. This arises from the lengths of objects in S_B (such as measuring rods) contracting in the direction of their motion according to:

$$L_B = \gamma L_\alpha$$

This result is completely consistent with the geometric method described previously.

Machian Simultaneity

It can be shown that the simultaneity of clock readings in each moving system S_A and S_B is different as it moves relative to the stationary system S_α. If the origins of the three systems are defined to be $t = 0$ when they coincide, the instantaneous clock readings along the x–axes of the two moving systems are given by:

$$t_B = -v_\alpha x_\alpha \, / \, c^2\sqrt{(1 - v_\alpha^2/c^2)} \qquad t_A = v_\alpha x_\alpha \, / \, c^2\sqrt{(1 - v_\alpha^2/c^2)}$$

These differences in clock readings of the moving systems are maintained relative to their respective origins, and therefore move with each moving system. This ensures that all moving clocks maintain their dilated clock rates relative to the stationary reference system according to the previous time dilation equations.

Consider in the rest system S_α two simultaneous events at the instant $t_\alpha 1$, for two locations on the x–axis $x_\alpha 1$ and $x_\alpha 2$. For these two points let $t_\alpha 1 = 0$ (this is the instant when the origins of S_α and S_B coincide). As a further simplification, let $x_\alpha 1 = 0$, which is by definition the origin of S_α. From the Machian Lorentz Transform Equations, we have:

$$t_B = (t_\alpha - x_\alpha v_\alpha/c^2) \, / \, \sqrt{(1 - v_\alpha^2/c^2)}$$

For $(x_\alpha 1, t_\alpha 1) = (0, 0)$ we have:

$$t_B 1 = (0 - 0 v_\alpha/c^2) \, / \, \sqrt{(1 - v_\alpha^2/c^2)} = 0$$

Because the origins coincide, this corresponds to a clock in S_B at $x_B = 0$ (the origin O_B), which reads $t_B = 0$. In other words, the clock at the origin of the moving system is synchronous with the corresponding origin–clock in the stationary system when they spatially coincide.

For $(x_\alpha, t_\alpha) = (x_\alpha 2, 0)$ we have:

$$t_B 2 = (0 - x_\alpha 2 v_\alpha/c^2) \, / \, \sqrt{(1 - v_\alpha^2/c^2)}$$
$$= - x_\alpha 2 v_\alpha/c^2 \, / \, \sqrt{(1 - v_\alpha^2/c^2)}$$

That is, the clock in S_B adjacent to the clock in S_α at $x_\alpha 2$ does not read $t_B 2 = 0$, and is therefore not synchronous with the other clocks.

GAMMA II
The Machian Lorentz Anti–Transformation

To summarise, the Lorentz transformations for each system are as follows:

$$x_B = (x_\alpha - v_\alpha t_\alpha) / \gamma \qquad\qquad t_B = (t_\alpha - x_\alpha v_\alpha/c^2) / \gamma$$

$$x_\alpha = (x_B + v_\alpha t_B) / \gamma \qquad\qquad t_\alpha = (t_B + x_B v_\alpha/c^2) / \gamma$$

These equations allow the space and time coordinates in one system to be calculated from those of the other. They therefore represent measurements that are objectively true for observers in either system – what these equations calculate for a system is true for observers in either system. The previous section proves that S_A and S_B are physically changed relative to S_α. This must be so, because light–spheres from S_A and S_B expand uniformly relative to S_α.

The two equations for x_α and t_α can be considered to be "anti–transformation" equations, but they have the same form as the transform equations for x_B and t_B. In other words, both sets of equations are covariant. However, it does not follow that the resulting space–time equations have to be covariant also, and the following method avoids this (and the resulting paradoxes).

Machian Anti–Dilation:
Explaining the Asymmetry

In order to calculate how clock rates in S_α vary relative to those in S_B, the Lorentz Transform for t_α is used:

$$t_\alpha = (t_B + x_B v_\alpha/c^2) / \sqrt{(1 - v_\alpha^2/c^2)}$$

If $T_\alpha = t_\alpha 2 - t_\alpha 1$ and $T_B = t_B 2 - t_B 1$, then we have:

$$T_\alpha = (t_B 2 + x_B 2 v_\alpha/c^2) / \gamma - (t_B 1 + x_B 1 v_\alpha/c^2) / \gamma$$

Now, because of the simultaneity changes in S_B, which are objectively true for observers in S_B and S_α, observers in S_B <u>must use the same clock arrangement</u> as observers in S_α (when determining S_B's time dilation). That is, to factor out Relativistic Simultaneity, only one clock in S_B is compared to two clocks in S_α. This is the <u>Machian Asymmetry Theorem</u>. This one clock in S_B has only one position, so $x_B2 = x_B1$ and the above equation therefore becomes:

$$T_\alpha = t_B2 / \gamma - t_B1 / \gamma = T_B / \gamma$$

That is, the measurement made by S_α is greater than the corresponding measurement made by S_B, which means S_α's clocks <u>run faster</u> than S_B's. This result for anti–dilation is completely algebraically consistent with the previous result for time dilation. That is, if $T_\alpha = T_B / \gamma$ then $T_B = \gamma T_\alpha$. It is also consistent with the fact that clocks in S_B have undergone an objective physical slowing due to their motion relative to the mutual centre of gravity of S_A and S_B.

Corollary

If instead of this, one clock in S_α is compared to two clocks in S_B, then we have $x_\alpha2 = x_\alpha1$. This gives:

$$\begin{aligned}
T_B &= (t_\alpha2 - t_\alpha1) / \gamma - v_\alpha (x_\alpha2 - x_\alpha1) / c^2\gamma \\
&= (t_\alpha2 - t_\alpha1) / \gamma \\
T_\alpha &= \gamma T_B
\end{aligned}$$

This is covariant with $T_B = \gamma T_\alpha$ and is superficially the same as the result in Special Relativity. However, whereas the result in Special Relativity is due to S_B being at rest, this equation in Machian Relativity represents an apparent dilation effect arising from simultaneity differences between the two clocks in S_B. This is because S_B has absolute motion and is physically changed due to its motion relative to S_α. This follows because its light–sphere is centered on S_α. This is consistent with De Sitter's Binary Star Principle, which proves that the motion of light is independent of its source motion and is relative to a Universal Reference Frame.

Because two clocks are used in S_B, this gives "simultaneity–apparent clock rates" which mimic the results for observers in the stationary system. This is not a problem because both systems can then use the previous method of two clocks in S_α to calculate the true changes in the clock rates for either moving system. It should also be remembered that "time–delay apparent dilations" occur when a receding clock is observed or compared to a single reference clock, and so different apparent effects occur which need to be allowed for.

Machian Anti–Contraction

In order to calculate how length measurements in S_α vary relative to those in S_B, the Lorentz Transform for x_α is used:

$$x_\alpha = (x_B + v_\alpha t_B) / \sqrt{(1 - v_\alpha^2/c^2)}$$

If $X_\alpha = x_\alpha 2 - x_\alpha 1$ and $X_B = x_B 2 - x_B 1$, then we have:

$$X_\alpha = (x_B 2 + v_\alpha t_B 2) / \gamma - (x_B 1 + v_\alpha t_B 1) / \gamma$$

$$= (x_B 2 - x_B 1) / \gamma - (v_\alpha t_B 1 - v_\alpha t_B 2) / \gamma$$

Because of the simultaneity changes in S_B, $t_B 2 \neq t_B 1$, and instead we must use $t_\alpha 2 = t_\alpha 1$, giving:

$$(t_B 2 + x_B 2 v_\alpha/c^2) = (t_B 1 + x_B 1 v_\alpha/c^2)$$

$$t_B 1 - t_B 2 = x_B 2 v_\alpha/c^2 - x_B 1 v_\alpha/c^2$$

Therefore the equation for X_α becomes:

$$X_\alpha = (x_B 2 - x_B 1) / \gamma - (x_B 2 v_\alpha^2/c^2 - x_B 1 v_\alpha^2/c^2) / \gamma$$
$$= (x_B 2 - x_B 1)(1 - v^2/c^2) / \gamma$$
$$= \gamma X_B$$

That is, when S_B measures a length, the corresponding length measured by S_α is less. From this, both systems conclude that objects in S_α have expanded according to:

$$L_\alpha = L_B / \gamma$$

This result for the Lorentz Anti–Contraction of S_α is algebraically consistent with the Lorentz Contraction for S_B, and results from the physical contraction of S_B as it moves relative to the mutual centre of gravity of S_A and S_B. The change in S_B measured by S_α is *absolute* because the cause of the change is inherent to S_B, which means the opposite change in S_α relative to S_B is *Relative* because the cause is external to S_α. The system S_α is unchanged and is at absolute rest because it still measures light from the other two systems to have spherical wave fronts.

Machian Anti–Simultaneity

The two inertial systems S_α and S_B are in motion relative to each other, but S_B (like S_A) has absolute motion, while S_α is at absolute rest at the centre of gravity of S_A and S_B. The Relativistic Simultaneity for observers in S_α has been determined, and this must now be done for S_B. Consider in the moving system S_B two simultaneous events at the instant t_B1, for two locations on the x–axis x_B1 and x_B2. For these two points let $t_B1 = 0$, which is the instant when the origins of S_α and S_B coincide. As a further simplification, let $x_B1 = 0$, which is by definition the origin of S_B. From the Machian Lorentz Transform Equations, we have:

$$t_\alpha = (t_B + x_B v_\alpha/c^2) / \sqrt{(1 - v_\alpha^2/c^2)}$$

For $(x_B1, t_B1) = (0, 0)$ we have:

$$t_\alpha 1 = (0 + 0 v_\alpha/c^2) / \sqrt{(1 - v_\alpha^2/c^2)} = 0$$

Because the origins coincide, this corresponds to a clock in S_α at $x_\alpha = 0$ (the origin O_α), which reads $t_\alpha 1 = 0$. This is consistent with the previous result for simultaneity in S_α.

For $(x_B2, t_B2) = (x_B2, 0)$ we have:

$$t_\alpha 2 = (0 + x_B 2 v_\alpha/c^2) / \sqrt{(1 - v_\alpha^2/c^2)}$$
$$= x_B 2 v_\alpha/c^2 / \sqrt{(1 - v_\alpha^2/c^2)}$$

That is, the clock in S_α adjacent to the clock in S_B at $x_B 2$ does not read $t_\alpha 2 = 0$, but its reading is advanced, and is therefore not synchronous with the other three clocks. However, unlike our previous results for simultaneity which are absolute, this result for anti–simultaneity is relative.

Another way is to consider when the origins are coincident as the systems pass each other. We know that clock readings on the positive x–axis of S_B are delayed, according to their position. So, to get such a clock to read $t_B = 0$, the system S_B needs to travel for a certain distance and duration. Because the earlier equation for simultaneity means S_B's clock reading is $t_B 2 = -x_\alpha 2 v_\alpha/c^2 / \sqrt{(1 - v_\alpha^2/c^2)}$, the system S_B needs to travel for an interval of $x_\alpha 2 v_\alpha/c^2 / \sqrt{(1 - v_\alpha^2/c^2)}$ so that the reading of this clock in S_B can advance sufficiently so that it then reads $t_B 2 = 0$.

Because the clocks in S_B are time dilated, this duration corresponds to an anti–dilation in S_α corresponding to $x_\alpha 2 v_\alpha/c^2 / (1 - v_\alpha^2/c^2)$. We also know that when comparing measurements, if S_α measures a length x_α, then S_B measures the same thing to have a length $x_B = x_\alpha / \sqrt{(1 - v_\alpha^2/c^2)}$. That is, the value measured by S_B is greater because its system has physically contracted, and so has shorter rulers. Therefore, we can also say that $x_\alpha = x_B \sqrt{(1 - v_\alpha^2/c^2)}$. This means that we can now express the anti–dilation of clocks in S_α to be:

$$T_\alpha = x_\alpha 2 v_\alpha/c^2 / (1 - v_\alpha^2/c^2)$$
$$= x_B 2 \sqrt{(1 - v_\alpha^2/c^2)} \, v_\alpha/c^2 / (1 - v_\alpha^2/c^2)$$
$$= x_B 2 \, v_\alpha/c^2 / \sqrt{(1 - v_\alpha^2/c^2)}$$

Superficially, this equation is covariant with the corresponding one for S_B. However, this covariance is *apparent*. The anti–simultaneity shift for S_α is *relative*, and can only be achieved in two instants – when the origins of S_α and S_B coincide, and when S_B has moved a certain distance to allow clocks on its x–axis sufficient time to advance their delayed readings to $t_B = 0$. However, the simultaneity shift for S_B is *absolute*, and occurs in one instant – when the origins of both systems coincide. This is because the delay in the readings on the positive x–axis of S_B is physically real, and light travels with absolute spherical motion relative to S_α. The delay ensures that light travelling in one direction along the axis of S_B has the value c as a constant speed. Relativistic Simultaneity is objectively true for observers in S_B – when they measure the same ray of light as observers in S_α then S_B travels an extra further distance before the light ray reaches the corresponding clock, whose reading has subsequently become synchronous – in an apparent sense. That is, other clocks in S_B have advanced by the same amount. Thus they remain asynchronous with each other, but all clocks in S_α remain synchronous with each other.

Corollary

A further result is that if observers in S_B make a "synchronous" measurement of a length, say x_B, then the corresponding length measured by observers in S_α is $x_\alpha = x_B / \gamma$. This might lead us to the conclusion that the greater reading is due to rulers in S_α shrinking due to motion relative to S_B. That is $L_\alpha = \gamma L_B$, and that the reciprocal changes predicted by Special Relativity are correct after all. However, the synchronous measurement made in S_B is apparent, and so is the resulting conclusion for the "contraction" of rulers in S_α. This is because the clock at the far end of the ruler in S_B is delayed due to Relativistic Simultaneity, and when it advances to read $t_B = 0$ the ruler has travelled further relative to S_α, and so the reading for the length measurement along the x–axis of S_α is correspondingly greater. Thus, Machian Relativity proves that covariance is apparent in nature. Rulers in S_B are physically contracted, so a gap appears between the two origins for the clock reading on S_B's axis to advance. So the two situations are not reciprocal.

Corollary – How Relativistic Simultaneity Causes Apparent Covariance

The apparent nature of the covariant dilation of clocks in S_α relative to S_B can be derived directly from the Relativistic De–Synchronisation of clocks in S_B. Consider two clocks in S_B being compared to one clock on the x–axis of the rest system S_α. From the equations for Relativistic Simultaneity, when the origins of systems S_B and S_α coincide, the reading of the clock in S_B corresponding to $x_\alpha2$ on the S_α x–axis is:

$$t_B2 = -v_\alpha x_\alpha2 \ / \ c^2 \ \sqrt{(1 - v_\alpha^2/c^2)} = -v_\alpha x_\alpha2 \ / \ c^2\gamma$$

The stationary clock in S_α at $x_\alpha2$ reads $t_\alpha2 = 0$. The corresponding reading for the clock at the moving origin O_B is $t_B1 = 0$.

When this moving clock O_B reaches $x_\alpha2$, the stationary clock at $x_\alpha2$ now reads $t_\alpha2 = T_\alpha = x_\alpha2 \ / \ v_\alpha$ (relative to S_B, the stationary clock $x_\alpha2$ undergoes relative to motion in the opposite direction to O_B). But because of Time Dilation, the moving clock at O_B reads $\gamma t_\alpha2 = \gamma T_\alpha$. Therefore the duration measured by the moving system is:

$$T_B = \gamma \ t_\alpha2 - t_B2 = \gamma \ T_\alpha + v_\alpha \ x_\alpha2 \ /c^2\gamma = \gamma \ T_\alpha + v_\alpha^2 T_\alpha \ /c^2\gamma$$
$$= T_\alpha \ (\gamma + v^2/c^2\gamma) = T_\alpha \ (\gamma^2 + v^2/c^2) \ / \ \gamma$$
$$= T_\alpha \ (1 - v^2/c^2 + v^2/c^2) \ / \ \gamma = T_\alpha \ / \ \gamma$$

Or: $\quad T_\alpha = \gamma \ T_B$

Thus the above result apparently predicts that relative to observers in the stationary system S_α, clocks in the moving system S_B have a faster rate (contradicting the expected Dilation) – while relative to observers in S_B, clocks in the stationary system S_α have a slower rate. In other words, we have arrived at the notorious Clock Paradox of Einstein's Relativity, with its symmetric (reciprocal) time dilations. However, in Machian Relativity, this effect is apparent, due to the physical and objectively real Relativistic Simultaneity of clocks in the moving system.

This proves beyond all doubt that comparing one clock in the stationary system S_α with two clocks in the moving system S_B produces apparent covariance effects that arise from the Relativistic Desynchronisation of clocks in the moving system – and that this effect in no way contradicts the true and objective Time Dilation for clocks in the moving system S_B. In other words, the physical desynchronisation of clocks in either moving system (S_A or S_B) generates an apparent symmetry with the stationary system S_α which does not imply that observers in either S_B (or S_A) can consider themselves at rest. This is the <u>Machian Apparent Covariance Theorem</u>.

Consider also time delays. For clocks in S_α at x_α and $-x_\alpha$, an observer at O_α sees both with an equal time delay x_α / c (and with equal readings). But for clocks in S_B at x_B and $-x_B$, an observer at O_B sees them with unequal time delays. This is because light spheres expand uniformly relative to S_α but not S_B (and these readings will be additionally different in S_B due to relativistic simultaneity). And as stated previously, for S_B to achieve simultaneous clock readings, S_B must travel a certain distance relative to S_α. Thus there is now a gap between O_α and O_B due to the physical length contraction in S_B.

GAMMA III
Relative Velocity in Machian Relativity

In all the previous equations, the speed v_α has been used for both systems. This represents the absolute speed of the moving system S_B relative to the stationary system S_α, as determined by observers in the system S_α. The question then arises, what is the relative velocity of the stationary system as judged by observers in the moving system? We might guess that it has the same numerical value v_α, but this needs to be proved.

Consider the two systems S_α and S_B as before, passing each other during relative inertial motion. When the origins of both systems coincide, the clocks at O_α and O_B both read $t_\alpha 1 = t_B 1 = 0$, as in the other examples. All of the other clocks in S_α are synchronised to read $t_\alpha = 0$, as this system is at absolute rest. Because the system S_B has absolute motion, its clocks are synchronised to have Relativistic Simultaneity to ensure a constant speed c for light along its axis of motion. In S_B, let there be a clock on the negative x–axis at $x_B = -x_B3$, which corresponds to $-x_\alpha3$ on the negative x–axis of S_α. Because of Relativistic Simultaneity, the initial reading on this clock is $t_B3 = \tau_B \neq 0$. S_B continues to move relative to S_α, so that its origin O_B reaches $x_\alpha3$ at $t_\alpha3$. At the instant $t_\alpha3$ for S_α, the clock in S_B at $-x_B3$ is coincident with O_α ($x_\alpha = 0$). Because of time dilation and relativistic simultaneity effects in S_B, the reading of the clock at $-x_B2$ is now $t_B3 = \tau_B + \gamma t_\alpha3$. The calculation follows:

For S_α the absolute velocity of S_B is v_α:

$$v_\alpha = (x_\alpha3 - x_\alpha1) / (t_\alpha3 - t_\alpha1) = x_\alpha3 / t_\alpha3$$

For S_B, let the relative velocity of S_α be v_B:

$$v_B = (x_B3 - x_B1) / (t_B3 - t_B1) = x_B3 / t_B3$$

The distance x_B3 travelled by O_α relative to S_B is numerically greater due to Lorentz Contraction within S_B and is in the opposite direction:

$$x_B3 = -x_\alpha3 / \gamma$$

The elapsed duration $t_\alpha2$ is less for S_B due to time dilation:

$$T_B = \gamma \, t_\alpha3$$

For the coincidence of O_α and O_B, we have $t_\alpha1 = t_B1 = 0$. Using the equations for Relativistic Simultaneity, the initial reading on the clock at $-x_B2$ is given by:

$$t_B3 = \tau_B = -x_\alpha3 \, (-v_\alpha / c^2) / \gamma = x_\alpha3 v_\alpha / \gamma c^2$$

That is, the reading is advanced to ensure the speed of light for S_B is constant in this direction.

For the later time $t_\alpha = t_\alpha 3$, the dilated duration for S_B is T_B, so the reading on the clock at $-x_B 3$ ($x_\alpha = 0$, when O_α reaches it) is now given by:

$$t_B 3 = \tau_B + T_B = x_\alpha 3 v_\alpha / \gamma c^2 + \gamma t_\alpha 3$$
$$= (x_\alpha 3 v_\alpha + \gamma^2 c^2 t_\alpha 3) / \gamma c^2 = (t_\alpha 3 v_\alpha^2 + \gamma^2 c^2 t_\alpha 3) / \gamma c^2$$
$$= t_\alpha 3 (v_\alpha^2 + \gamma^2 c^2) / \gamma c^2 = t_\alpha 3 . c^2 / \gamma c^2$$
$$= t_\alpha 3 / \gamma$$

Therefore:

$$v_B = x_B 3 / t_B 3 = -(x_\alpha 3 / \gamma) \times (\gamma / t_\alpha 3)$$
$$= x_\alpha 3 / t_\alpha 3 = -v_\alpha$$

That is, the velocity of S_α relative to S_B is equal and opposite to the velocity of S_B relative to S_α. This means the relative speeds are the same: $|v_B| = |v_\alpha|$. However, unlike in Special Relativity, this does not imply complete equivalence (covariance for time dilation and length contraction), as v_B is relative, while v_α is absolute (e.g. S_B moves relative to the average momentum of the whole system while S_α does not).

Corollary

Similarly, it can be shown that relative velocities do not combine in the usual way. Relative to the system S_α let the velocity of S_A be v_α, and let the velocity of S_B be $-u_\alpha$. It can be shown that the relative velocity between S_A and S_B is given by:

$$w = (v_\alpha + u_\alpha) / (1 + v_\alpha . u_\alpha / c^2)$$

That is, the relative velocity between S_A and S_B is not $w = v_\alpha + u_\alpha$ as expected by Galilean Relativity. This result is the same as for Special Relativity, and together with mass increase means that relative velocities cannot reach or exceed the speed of light. For low absolute velocities, the result approximates to that of Galilean relativity.

Corollary – Light Spheres and Solid Spheres

In Machian Relativity, the wavefronts from a source have absolute motion relative to the average momentum of matter (due to the Mach Field interactions of that matter). Thus, for observers at absolute rest, the centres of light spheres remain at absolute rest, *regardless of the source motion*. And for observers moving with the source, the centre of the light sphere recedes. The observers' measuring rods are contracted in the direction of their absolute motion, so the light sphere appears extended along the axis of its relative motion (Appendix IV). This is the Machian Wavefront Theorem. For a sphere of solid matter with absolute motion, it is physically contracted in the direction of its motion. So relative to observers at absolute rest, the sphere is flattened along this axis (oblate spheroid). However, observers moving with the sphere detect no change in its shape. This is because their measuring rods are correspondingly contracted. Also, this means that stationary solid spheres are elongated along their motion relative to moving observers (prolate spheroid). This applies to any form of motion.

Conclusion

According to Kant, we must distinguish between things as we perceive them (phenomena) and as they really are (noumena). The same applies in Relativity – the apparent effects of Time Delays need to be factored out, to allow the true dilation to be determined. And the covariance due to relativistic simultaneity needs to be factored out, so that absolute changes can be distinguished from relative ones, and hence absolute motion from relative motion[2].

---

[2] The existence of anti-dilation (inverse or reverse dilation) has been experimentally demonstrated in the Hafele-Keating Experiment, where a clock flown east-west (against the earth's rotation) has an increased rate. Thus the assumption that there can only be absolute time between moving frames (Newtonian non-dilation), or relative time (Einsteinian reciprocal dilation), is a *false dichotomy*. Rather, there is both absolute and relative time between such frames (Machian physical dilation).

Glyn Phillips

DELTA

THE EQUATIONS OF MACHIAN RELATIVITY DERIVED BY GEOMETRIC METHODS

Previously, the equations of Machian Relativity have been derived using the same algebraic methods initially developed for Special Relativity. In order to gain a better understanding of the equations, they will now be derived according to geometric principles.

The Absolute Motion of Light

Consider the two identical systems S_A and S_B moving inertially as described in the previous examples. Each system has its own light source. As before, both systems emit spherical light pulses when their origins pass each other. Let the wavefront of the light pulse from S_A be Ψ_A, and let the wavefront of S_B's pulse be Ψ_B.

Each system has its own Mach Field, Φ_A and Φ_B respectively. The wavefronts move through both Mach Fields, and are equally dragged in the direction of either system. These dragging effects cancel each other, and thus the wavefronts from both sources remain centered on the centre of gravity of both systems, which is the mid–point. A third reference system S_α is placed with its origin at rest at the centre of gravity. Observers in this third system determine the two wavefronts to remain centered on its origin. For observers in the other two systems, the wavefront centres recede from their respective origins. The third system is placed at the centre of gravity of the other two. Thus the average momentum remains unchanged, and it has no overall effect on the other two.

Machian Time Dilation

Systems S_A and S_B have the same motion $v/2 = v_\alpha$ relative to their common centre of gravity. Therefore, any solution for S_A must also

51

apply equally to S_B. The third identical system S_α is placed at the centre of gravity of S_A and S_B. As shown in the previous chapter, light from both sources in S_A and S_B moves absolutely relative to S_α. For a diagonal light beam in S_α, light moves up the y–axis of system S_B (y_B):

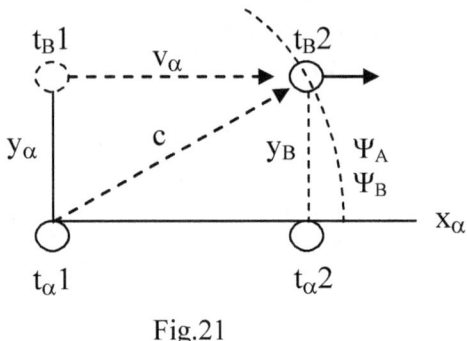

Fig.21

The clocks at the origins of both systems are synchronised to read $t_\alpha 1 = t_B 1 = 0$. At time $t_\alpha 2$ we have for durations in each system:

$$t_\alpha 2 - t_\alpha 1 = T_\alpha \qquad\qquad t_B 2 - t_B 1 = T_B$$

Let the speed of light c be constant for both systems. Therefore:

$$y_\alpha^2 = c^2 T_\alpha^2 - v_\alpha^2 T_\alpha^2 \qquad \text{(For } S_\alpha - \text{ by Pythagoras's}$$
Theorem)
$$y_B^2 = c^2 T_B^2 \qquad\qquad \text{(For } S_B)$$

It is assumed that lengths along the y–axis remain unchanged for both systems, so that $y_\alpha^2 = y_B^2$. The above equation therefore becomes:

$$c^2 T_\alpha^2 - v_\alpha^2 T_\alpha^2 = c^2 T_B^2.$$

From this, we have:

$$T_B = T_\alpha \sqrt{(1 - v_\alpha^2/c^2)}$$

That is, observers in <u>both systems</u> measure the reading on S_B's clock to be lower than S_α's clock, and therefore observers in both systems conclude that S_B's clocks are working at a slower rate than S_α's clocks. This is known as time dilation. The same reasoning also applies to clocks in system S_A. This method is identical to the Light Clock method described in many books.

When S_B is initially at rest, light beam from its source (and the other) is diagonal. But when S_B has the appropriate absolute velocity, the same light beam now travels along its y–axis. This is an example of aberration.

These are proper physical changes, which not only obey the rules of logic, but of algebra also, so that we can also write:

$$T_\alpha = T_B / \sqrt{(1 - v_\alpha^2/c^2)}$$

That is, if S_B's clocks go slower relative to S_α's, then according to logic we expect that relative to S_B, S_α's clocks have a faster rate, and the algebraic manipulation shown above demonstrates this to be the case. There is no corresponding covariant equation. This is because of the Mach Field interactions between S_A and S_B, so light spheres from moving sources remain centred on S_α but not on S_B.

Machian Length Contraction

As described previously, systems S_A and S_B are moving inertially, with the third system S_α remaining at rest at the common centre of gravity of S_A and S_B. In the previous example for deriving time dilation, the motion of a diagonal light beam in the system S_α was considered. In this example, we now consider the part of the spherical pulse that corresponds to a light beam parallel to the x–axes of both systems. Because light moves absolutely relative to S_α, this light can come from either source in S_A or S_B.

Consider the system S_B to carry a rigid rod parallel to the direction of motion, determined by S_α to have length L_B. When S_B is initially at rest relative to S_α, we have $L_B = L_\alpha = L_0$, which ensures both systems measure the speed of light to be the same, because their clocks are also identical. Observers in the system S_B consider light (from either source) that moves from the origin O_B of S_B (when it also initially coincides with O_α) to the end of the rod, and is then reflected back to the origin O_B. The round trip light motion brings light back to the same clock at S_B's origin. This is necessary to avoid simultaneity changes in readings of different clocks along the x–axis of S_B. The outward journey to the far end of the moving rod is:

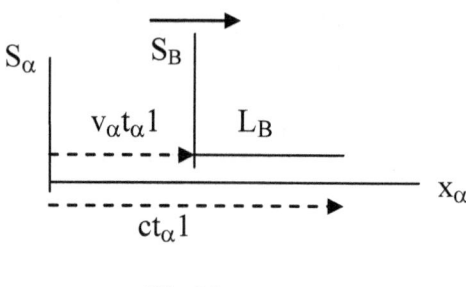

Fig.22

Relative to the system S_α, the distance for the outward journey of the light beam to the end of the rod is given by:

$$x_\alpha 1 = ct_\alpha 1 = L_B + v_\alpha t_\alpha 1$$

$$t_\alpha 1 = L_B / (c - v_\alpha)$$

The return journey to the moving origin is:

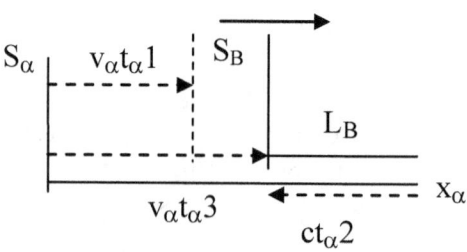

Fig.23

The distance for the return journey of the same beam from the end of the rod back to the origin of S_B, is determined by S_α to be:

$$x_\alpha 2 = -ct_\alpha 2$$

The term "–c" is used because the light beam has had its motion reversed. The total distance travelled by the light beam along the x–axis of S_α is therefore:

$$x_\alpha 3 = x_\alpha 1 + x_\alpha 2 = v_\alpha t_\alpha 1 + L_B - ct_\alpha 2$$

The total time for whole journey is:

$$t_\alpha 3 = t_\alpha 1 + t_\alpha 2$$

In the time $t_\alpha 3$ made by the light beam for the round trip journey back to the origin of S_B, the distance travelled by O_B relative to S_α is given by:

$$x_\alpha 4 = v_\alpha t_\alpha 3 = x_\alpha 3$$

The previous equation then becomes:

$$v_\alpha t_\alpha 1 + L_B - c\,(t_\alpha 3 - t_\alpha 1) = v_\alpha t_\alpha 3$$

Next, we consider the diagonal light ray in S_α previously used to determine S_B's time dilation. For observers in S_B, this light ray travels up the axis y_B. If it is reflected back down this axis to the origin O_B, c–invariance requires its arrival to coincide with the return of the light wave reflected along the axis x_B. There are no relativistic changes in lengths along the y–axis, thus $y_\alpha = y_B$. The path of the transverse beam in the moving system is:

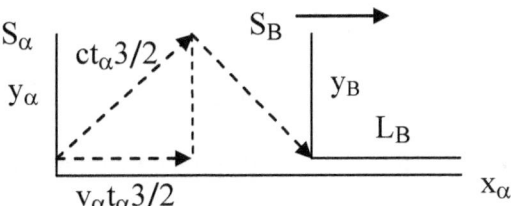

Fig.24

Therefore, by Pythagoras's Theorem:

$$y_\alpha^2 = c^2 (t_\alpha 3/2)^2 - v_\alpha^2 (t_\alpha 3/2)^2$$

Rearranging the above equation gives:

$$t_\alpha 3 = 2y_\alpha / \sqrt{(c^2 - v_\alpha^2)}$$

Combining these equations gives:

$$v_\alpha L_B / (c - v_\alpha) + L_B - c \left[2y_\alpha / \sqrt{(c^2 - v_\alpha^2)} - L_B / (c - v_\alpha) \right]$$
$$= v_\alpha 2y_\alpha / \sqrt{(c^2 - v_\alpha^2)}$$

Solving this equation gives:

$$L_B = y_\alpha \sqrt{(1 - v_\alpha^2/c^2)}$$

That is, as the motion v_α of S_B relative to S_α increases, the lengths of objects in S_B decrease relative to S_α in the direction of their motion. In addition, we must have $y_\alpha = L_\alpha = L_0$, because S_α measures the speed of light to be the same in all directions. This gives:

$$L_B = L_0 \sqrt{(1 - v_\alpha^2/c^2)}$$

That is, the rod in S_B contracts due to its motion relative to the centre of gravity of S_A and S_B (in fact, the average momentum of both systems) – even when the reference system S_α is not present. Similarly, rods in S_A change in exactly the same way:

$$L_A = L_0 \sqrt{(1 - v_\alpha^2/c^2)}$$

Corollary I

The change in L_B is a proper physical change which is objectively true for S_α, S_A and S_B. This is exactly the same as the objective time dilations for clocks in S_A and S_B. The Lorentz Contraction of L_B must also be true for observers in S_B, because if it was not, L_B

would not be contracted for them, and they would measure a variable speed of light according to Classical Relativity. The fact that they measure light reflected along the x–axis to have a constant speed means they conclude that their rulers have physically contracted. This is a normal physical change that obeys the rules of logic and algebra. That is, their conclusions are consistent with other observations they make. For example, they measure the lengths of objects in S_α to *expand* (Lorentz Anti–contraction), and they also observe light from their source in O_B to be offset from O_B, and instead centred on O_α. So, the *motion* of this light is absolute relative to O_α, even though the *speed* of the light is constant (absolute) relative to S_B.

Corollary II

From this change, the relationship between the measurements made by both systems can be deduced. If S_α measures an object to have a length X_α on its x–axis, then the corresponding measurement of the same length made by S_B will be:

$$X_B = X_\alpha / \sqrt{(1 - v_\alpha^2/c^2)}$$

The greater magnitude in the measurement made by S_B arises from the fact that its rulers are physically shortened by the Lorentz contraction. Therefore, relative to S_B, lengths in S_α increase according to the equation:

$$L_\alpha = L_0 / \sqrt{(1 - v_\alpha^2/c^2)}$$

Where, L_0 is the length of a rod at rest in S_B as measured by its observers. The fact that observers in S_B measure their rod to have a length L_0 does not imply that they can regard themselves "at rest", or that the equations have to be symmetric ("covariant"). What it means is that all objects in S_B have been equally contracted so there is no relative change between them (there is, however, an absolute change).

Machian Simultaneity

So far, the equations for time dilation and Lorentz contraction have been derived. Time dilation ensures that the speed of light along the y–axis of the moving frame is constant, while Lorentz contraction ensures that the speed of light along the x–axis of the moving system is constant, for a round trip between the system origin and the end of the rod. However, the speed of light for a light beam moving in one direction only along the x–axis of the moving system now needs to be considered.

When the origins of both systems coincide, clocks at the origins of both systems both read $t_\alpha 1 = t_B 1 = 0$. Lengths of objects in the moving system S_B are physically contracted to $L_B = \gamma L_\alpha$ by motion relative to the stationary system S_α, which is at the centre of gravity of S_A and S_B. This is Machian Lorentz Contraction. Spherical light pulses are emitted from both origins, whose wavefronts have absolute motion relative to S_α. The light ray travels from both coincident origins, and travels to the end of the contracted rod as S_α and S_B diverge from each other. The motion of a light beam from one end of the moving rod to the other is therefore:

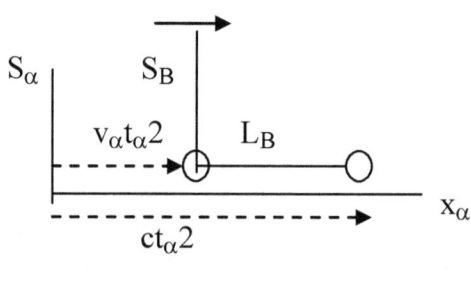

Fig.25

For S_α, in time $t_\alpha 2$ the system S_B moves a distance $v_\alpha t_\alpha 2$, so the distance moved by the light ray is:

$$ct_\alpha 2 = v_\alpha t_\alpha 2 + L_B = v_\alpha t_\alpha 2 + \gamma L_\alpha$$

So:

$$t_\alpha 2 = \gamma L_\alpha / (c - v_\alpha) = \gamma L_0 / (c - v_\alpha)$$

[We have $L_\alpha = L_0$ because S_α is at absolute rest at the centre of gravity of S_A and S_B]. The clock at the end of L_B reads $t_B 2$ and its speed of light is also constant, therefore:

$$c = (L_B - 0) / (t_B 2 - 0)$$
$$ct_B 2 = L_B = L_0$$

[Even though the rod L_B is contracted due to motion relative to S_α, the length of the rod in S_B for its observers is $L_B = L_0$. This is not because S_B is somehow "at rest", with S_α symmetrically contracting relative to S_B – as Einstein would have us believe – but because all lengths in S_B are equally contracted.]

When the origins of both systems originally coincided, the reading of the clock at the end of the moving rod is given by $t_B 1 = t_B 2 - \gamma\, t_\alpha 2$, where $\gamma\, t_\alpha 2$ is the time dilated rate of S_B's clocks due to motion relative to the stationary system S_α. Therefore:

$$ct_B 1 = ct_B 2 - \gamma\, ct_\alpha 2$$
$$= L_0 - \gamma\, ct_\alpha 2$$

Substituting the value for $t_\alpha 2$, this gives:

$$ct_B 1 = L_0 - c\, \gamma^2\, L_0 / (c - v_\alpha)$$

From this, it can be shown that:

$$t_B 1 = L_0 (-v_\alpha / c^2)$$

This is the reading of the clock at the end of the contracted moving rod (γL_0) when the origins of both systems initially coincided. This is required so that the moving system measures the speed of light to be c when the light ray from the origin reaches this clock at a later time. If instead we consider a rod initially of length L_0 / γ when at rest, this will have a length L_0 moving relative to S_α due to Lorentz Contraction, and the reading of a clock at its end (when both origins coincide) will (by scaling) now be:

$$t_B 1 = L_0 \left(-v_\alpha / \gamma c^2 \right)$$

This result is entirely consistent with the predictions of Special Relativity as derived from the Lorentz Transform Equations, and both results are quantitatively the same (but qualitatively different). Generalising the above equation for a row of moving clocks in S_B that correspond to any value of x_α (and not just $x_\alpha = L_0$) we have:

$$t_B(x_\alpha) = - x_\alpha . v_\alpha / \gamma c^2$$

This gives the readings for clocks in S_B when the origins of both systems initially coincide. It can be seen that unlike clocks in S_α, they are no longer synchronous. The particular synchronization they have is maintained relative to the origin of S_B, which ensures they all have the same time dilation as they move relative to S_α. Like time dilation and Lorentz contraction, this Relative Simultaneity is objectively true for S_α and S_B – so, *the clock readings predicted by this equation must also be true for observers in S_B*, because if it was not, observers in S_B would measure the speed of light to be variable according to Classical Relativity, which we know is false.

For example, a moving observer and a stationary observer at $x_\alpha > 0$ pass each other in the same vicinity. At this instant, if the stationary observer's clock reads $t_\alpha = 0$, the moving observer's clock reads $t_B(x_\alpha) < 0$, and *this is true for both observers*.

Machian Aberration

Consider the previous example for time dilation, but with some modifications. An observer is placed at the origin of S_B, and in S_α the clock at y is replaced with a point source that continuously emits spherical light pulses. The observer is assumed to have no significant size with respect to y_α, and each of his eyes approximates to a pinhole camera. When S_B is at rest with S_α, the light source appears along y_B relative to the observer, because y_B and y_α are coincident. The light beam travels through the pinhole to the back of the eye. However, if S_B has absolute velocity relative to S_α the light beam travels diagonally relative to the observer, as he looks towards the source at y_α. The beam therefore travels at an angle through the eye and so does not reach the back to form an image of the source. However, if the observer instead looks at an angle in the direction of his absolute motion, the beam is able to reach the back of his eye to form an image. Thus, the apparent position of the source is shifted from being perpendicular to the observer's motion to being slightly ahead in the direction of motion.

Now, if a second source instead travels with the observer, its spherical wavefronts must still expand uniformly relative to S_α. This is because S_B's light spheres are partially dragged in the opposite direction by S_A. Beams perpendicular to the motion cannot reach the observer. However, a beam emitted diagonally in the direction of absolute motion can reach the observer. If the observer is looking sideways to his motion, the beam can reach the back of the eye. So the observer is able to see the stationary source, but the light has been emitted in a different direction. Also, the observer can see more of this light if he looks at an appropriate angle behind the true position of the source.

Conclusions

These examples, together with others in this book, prove that the changes in relativity are physical in nature and therefore asymmetric – even for inertial motion. Statements that claim the equations of Special Relativity are "unphysical", "kinematic" or "how something looks relative to an observer" are false.

The changes in a system as seen by an observer in a second system are not just a relative effect for the observer, but a physical change in the system itself. They are deduced from the measurements made by observers in the first system itself. This ensures observers in the first system measure the same speed of light as those in the second, so that the change is objectively true for both sets of observers. This is the simplest explanation – the system looks different to the observer because it has undergone an inherent physical change.

This geometric method is independent of the algebraic method, yet its asymmetry is consistent with that of Machian Relativity. This method cannot produce the symmetry of Special Relativity, as the location of the wavefronts (from either source) is an objective fact determined by the relative masses of the systems, and cannot be arbitrarily shifted.

Consider the equation $c^2t^2 - x^2 = c^2t'^2 - x'^2 = 0$ for two relatively moving inertial systems. This means that the speed of light is constant in all directions for both systems, but it does not mean that light moves as an expanding sphere relative to each system. In other words, the *speed* of light is absolute (constant) for both systems but the *motion* of the light's wavefront is not. And the variables (x, t) and (x', t') <u>are not</u> how one system subjectively appears to the other due to their relative motion – they are the respective measurements made by observers in each system – which result from physical changes that are objectively true for both systems. So these equations represent how each system <u>actually is</u> – not how it looks to external observers [For this example, they are also how each system looks relative to the other because the systems are directly compared to avoid apparent effects].

Special Relativity is based on two postulates (or principles):

1. The Relativity Principle – The laws of physics are the same for all inertial systems – which means covariant equations and relative motion (no absolute motion for inertial frames).
2. The c–Invariance Principle – The speed of light is constant for all inertial systems, regardless of any relative source motion.

The problem lies with how the (Einsteinian) Relativity Principle is formulated. To resolve the issue, we have the <u>Principles of Machian Relativity</u> instead:

1. The Machian Relativity Principle – The laws of physics are the same for all inertial and accelerated systems – which means invariant equations and motion relative to the average momentum of all matter (inertial systems have absolute motion).
2. The c–Invariance Principle – The speed of light is constant for all inertial systems, regardless of any relative source motion.

Special Relativity has the following epistemological defect; the relativistic effects between two systems persist across space regardless of their separation, *without any causal explanation*. Hence Mach Fields must exist, and space is not a true void. Galileo showed that sideways and falling motions occur together (e.g. a parabola). This refuted Aristotle's idea that they occurred separately. And Mach Fields also produce average effects in combination, not separately (reciprocally) as in Special Relativity.

An observer in gravitational free-fall feels no acceleration effects, but does this mean he can conclude he is not accelerating, and is "at rest"? <u>No</u> - the higher an object is dropped from, the greater its impact. Similarly for the Michelson-Morley Experiment – does the null result mean the frame of the earth is at rest? Again, <u>No</u> – the time dilation of earth clocks in the Hafele-Keating Experiment proves the earth's frame is not at rest. And stellar aberration proves the earth is moving relative to the Universal Reference Frame for all light. The null result's *true meaning* is that the apparatus experiences physical contraction, to maintain light-speed constancy in the earth's frame.

EPSILON

BUILDING THE UNIVERSE:
THE GENERALISED PRINCIPLE OF MACHIAN
RELATIVITY ACCORDING TO THE METHOD OF MACH
FIELDS

So far, it has been shown that two identical systems undergo relativistic changes due to motion relative to their common centre of gravity, which is the geometric midpoint. That is, they do not change due to motion relative to each other (as required by Special Relativity). In Machian Relativity, the space-time equations describing a system with absolute motion (for observers at absolute rest) are identical to those in Special Relativity. But, the equations describing a system at absolute rest (relative to observers with absolute motion) have apparent covariance and physical asymmetry. Two further aspects will now be addressed. Firstly, the combined effects of multiple systems with different masses; and secondly, an extension to other relativistic processes other than those of space-time (clocks and rulers).

EPSILON I
The Combined Influence of All Matter in the Universe

If one of two systems has a greater mass, then its Mach Field (Φ) has a greater influence on the other system than vice versa. So, the less massive system therefore undergoes greater relativistic changes than the more massive system. The change in each system is determined by their speeds relative to their common centre of gravity, which is closer to the more massive system. For example, a system S_A twice as massive as a second system S_B has twice the influence on the second system than vice versa. So, for equal system masses, $m_A : m_B = 1 : 1$, the velocity ratio relative to the centre of mass are $v_A : v_B = 1 : 1$. But for $m_A : m_B = 2 : 1$, we have

$v_A : v_B = 1 : 2$. These absolute velocities determine the respective relativistic changes for each system.

The first system can be considered as being two systems each of the same mass as the second system. Thus, it has *twice* the influence on a single system. This can be determined from their effects on light spheres. The more massive system drags the light sphere more than the less massive, so that it remains centred on the centre of gravity. Thus, the motion of the more massive system relative to the centre of the light sphere (and the common centre of gravity) is reduced, and so its space-time changes less (to satisfy c-invariance).

Now, all masses in the universe have their own Mach Fields, and their effect on the first two systems also needs to be included. For example, if a third system S_C is introduced (Fig.26), then a new common centre of gravity is generated based on the relative masses of these three systems (m_A, m_B, m_C). Because Mach Fields are *spatially uniform*, the change is also relative to any fixed point in space from the centre of gravity. So it is better to say the change occurs due to speed relative to the *average momentum* of the systems (Γ_{ABC}):

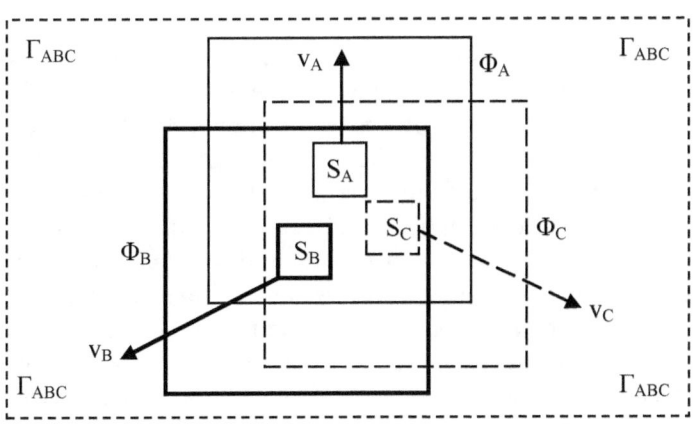

Fig.26: Mach Fields and Average Momentum

Each Mach Field (Φ_A, Φ_B and Φ_C) uniformly transmits its system's *causal effects* across space. So, each system relativistically changes due to its velocity relative to the new average momentum. And by including the influence of all other masses in the universe, it therefore follows that any system must change due to its velocity (v_p) relative to the average momentum of all matter in the universe (Γ_∞). Thus, the relativistic coefficient $\gamma = \sqrt{(1 - v_\alpha^2/c^2)}$ becomes $\gamma = \sqrt{(1 - v_p^2/c^2)}$. Space is therefore a <u>Plenum</u>, and is filled with the relatively moving Mach Fields (Relative Causal Spaces) of masses. And, this is a general law that *applies to any form of motion, inertial or accelerated.* So, Lorentz contraction, time dilation and simultaneity are determined by motion relative to the average momentum of all matter. This is Mach's Principle.

Scholium

The Postulate of Mach Fields is entirely consistent with other physical phenomena. In his Theory of Universal Gravitation, Newton postulated the existence of gravitational fields to explain the orbit of planets around the sun, the moon's orbit around the earth, and why objects fall to the ground. In the same way, I have postulated the existence of Mach Fields to explain relativistic processes.

Gravitational fields extend to infinite distance, and I say Mach Fields do the same. In fact, the Mach Field and can be considered to be a different aspect of the gravitational field, as both are associated with mass. The difference is that the force of gravity decreases with distance from a mass due to the inverse square law, while a Mach Field's relativistic effects are spatially uniform, e.g. the speed of a light ray is constant regardless of how far it travels from an observer or system. And Machian Relativity tells us that this arises in two ways – either the system is at absolute rest in the Plenum (unchanged space-time), or the system has absolute motion (changed space-time). Because the force of gravity decreases with distance from a source mass, its effects are localised. However, the spatial uniformity of Mach Fields means their effects are global (cumulative), and so they cannot be considered in isolation.

Corollary: The Earth Approximates to a State of Absolute Rest

There is no need to consider the mass and speed of every atom, planet and star in the universe to determine a body's absolute motion relative to all matter. Bodies with more mass have lower accelerations than less massive bodies when subjected to forces of identical magnitude. Therefore, massive bodies, such as stars and planets have lower absolute velocities than (say) subatomic particles. So, massive bodies (such as the earth) approximate to systems at absolute rest.

Corollary: The Plenum is Immovable

When two bodies interact, their shared momentum is preserved due to Newton's Third Law. Therefore, the average momentum in the universe is unchanged, and is a true state of absolute rest.

EPSILON II
Extending Mach's Principle to Other Phenomena

The Absolute Motion of Light

It has previously been shown that the centres of light spheres from sources in two relatively moving systems must remain at rest relative to the average momentum of both systems. By including the influence of masses in the rest of the universe (by their Mach Fields acting across space), it follows that the centres of light spheres must remain at rest (e.g. uniform expansion) relative to the average momentum of all matter. Therefore, all wavefronts travel at speed c relative to the average momentum of all matter, and this motion is absolute. The same wavefronts also travel at speed c relative to inertial systems moving relative to the Plenum, but this is because the space-time (clocks and rulers) of such systems have physically changed. This is the Machian Wavefront Theorem and applies to sources with any form of motion. The issue has never been addressed by conventional science. If scientists and authors wish to dispute this, they *need to give their own explanation.* But, light spheres (from accelerating sources) must still expand non-uniformly relative to their observers, so asymmetries cannot be circumvented.

Corollary I: Inertia

The same applies to inertia. Let us apply the Method of Mach Fields. As an initial assumption, let all matter in the Universe be assumed to obey Newton's First Law relative to the Mach Field of an arbitrarily chosen system (with mass), even when the system is accelerated. That is, when the system is accelerated, it mutually interacts with matter (such as rocket gases) according to Newton's Third Law. However, all other external matter in the universe shares in the system's acceleration, by virtue of its Mach Field, which uniformly fills all of space. This means observers in the accelerated system do not experience acceleration effects. This is of course false, but the Principle of Machian Relativity leads to a proper solution. By including the mass of a second system, its Mach Field counters the Mach Field of the first system, so that all external matter obeys Newton's First Law relative to the average momentum of both systems. Therefore, observers in the first system now experience acceleration effects. By including the influence of Mach Fields from all other masses in the rest of the universe (including the rocket gases), it follows that the inertia of all bodies must be relative to the average momentum of all matter. This explains why systems experience acceleration effects when subjected to applied forces. Thus, non-inertial effects arise due to Mach's Principle, and this provides a better explanation than either Einstein's Equivalence Principle (arbitrarily induced gravitational fields) or Newton's Absolute Space. So, Newton's First Law of Motion, according to Mach's Principle, is:

"A body remains at rest, or moves in a straight line at constant speed, relative to the Plenum, unless acted on by a resultant force".

Corollary II: Relativistic Mass-Energy Increase

Using similar reasoning, it follows that relativistic mass-energy increase must occur due to velocity relative to the average momentum of all matter (e.g. absolute motion). This approximates to velocity relative to the earth. This can be done by modifying the mass-energy equations of Special Relativity using the Method of Mach Fields [e.g. if two identical masses change in the same covariant way due to due motion relative to the other, then they both physically change relative to their centre of gravity, because of the self-influencing effects of their respective masses].

The resulting mass-energy equations are therefore <u>asymmetric</u>, because the equations are no longer based on relative velocity. For example, if two identical masses m_A and m_B are at absolute rest (relative to distant matter), and m_B is given an absolute velocity v_p, then $m_B = m_A / \sqrt{(1 - v_p^2/c^2)}$, which is an *absolute change*.

And relative to m_B, $m_A = m_B \sqrt{(1 - v_p^2/c^2)}$, which is a *relative change, due to the cause being within m_B itself.* Therefore, these equations obey the rules of algebra and logic.

Corollary III: Work Done by Forces

The same reasoning applies to work done by a force. If, according to conventional science, "everything is relative", then the work done on mass m_A (relative to m_B) is $W_A = m_A[a]_R \bullet [d]_R$ and relative to m_A we have $W_B = m_B.[a]_R \bullet [d]_R$ for the work done on m_B. In both equations $[a]_R$ and $[d]_R$ are vectors for the relative acceleration and distance moved between either mass.

However, each mass has a Mach Field. Thus, by the Method of Mach Fields, the work done on either body is now $W = m[a]_\Gamma \bullet [d]_\Gamma$ (where $[a]_\Gamma$ and $[d]_\Gamma$ are the acceleration and distance moved relative to the centre of gravity Γ of m_A and m_B). By including all other matter in the universe, the work done (on any mass m) is given by; $W = m[a]_P \bullet [d]_P$ (where $[a]_P$ and $[d]_P$ are the absolute acceleration and distance moved relative to the average momentum of all matter in the universe). This approximates to motion relative to the earth. This method is superior to conventional science, which uses Einstein's "induced gravitational fields" to make acceleration a reciprocal effect. Instead, work done is defined as a change in absolute motion due to the action of proper forces arising from mutual interactions between masses (e.g. Newton's

Third Law). Alternatively, consider a mass at rest in the Plenum. Its gravitational field does work on a photon according to:

$$\partial E = \text{force} \times \text{distance} = \partial/\partial t \, (mv_p) \times \partial x_p$$
$$= (m\partial v_p/\partial t + v_p\partial m/\partial t) \, \partial x_p = mv_p\partial v_p + v_p^2\partial m$$

For a photon, we have $v_p = c$ and $\partial v_p = 0$, so $\partial E = c^2\partial m$. Integration gives:

$$E = mc^2$$

Also if a photon is absorbed by a mass (m), we can write:

$$\partial E = c^2\partial m = mv_p\partial v_p + v_p^2\partial m$$

Thus: $m = m_0 / \sqrt{(1 - v_p^2/c^2)}$

Therefore changes in mass-energy are not due to changes in relative velocity, but *changes in absolute velocity*. Forces can increase absolute velocity and mass-energy, and they can similarly decrease these quantities. This applies to any form of motion. It also follows that acceleration asymmetries will arise within an inertial system with absolute motion. For example, if a body is accelerated in the direction of the system's absolute motion, its mass increase impedes acceleration. But if the acceleration is against the absolute motion, mass decrease will give a different acceleration relative to the system. This is because both outcomes are determined by motion relative to the External Universe, but not the system itself. This is not so strange, if we consider the asymmetries on the rotating earth (e.g. rocket launches).

Corollary IV: Electromagnetism

Similarly, the magnetism of electric currents is also determined by their absolute velocity (inertial or accelerated) relative to the average momentum of all matter. That is, currents move through the Mach Fields of all matter, and this resolves the Paradox of Magnetism.

Scholium

Einstein's idea of relativity based on changes due to relative motion has been demonstrated to be false. For example, when a train accelerates, it mutually interacts with the earth. But, we do not say "Observers both in the train and on the earth must experience the same acceleration effects, because they are both non-inertial". We know that is false, because the earth is more massive than the train, and so only the train experiences significant acceleration. The same reasoning applies to time dilation and inertial systems. It does not follow that two relatively moving inertial systems experience the same (reciprocal) time dilation because "They are both inertial". Rather, it is more consistent if the more massive system induces a greater dilation in the less massive system than vice versa.

But, it can be shown that the seeds of Machian Relativity are already contained in Special Relativity. Let me explain. If, according to Einstein's Relativity Postulate, two inertial systems change symmetrically due to their shared relative motion, then the change must be maintained regardless of their changing separation. Hence, each system influences the other across space, and this must be due to the existence of a Mach Field associated with each system. Now, a system's Mach Field arises from its mass. The other system's motion relative to the Mach Field of the first system induces its space-time properties (clocks and rulers) to change. But, the other system has its own associated mass, and this tends to oppose the changes due to the first system. Hence, the second system partially changes (relative to the shared average momentum). This is the Principle of Self-Influence. Likewise, the first system partially changes due to its own mass and the momentum of the second system. So, we now have a relativity theory founded on the principle of velocity and mass. And by extension, the influence of the entire external universe (Bishop Berkeley's "Fixed Stars") needs to be included. From this, the concepts of absolute motion and absolute rest are seen to be self-generating. Thus Machian Relativity has a proper cause-effect mechanism – relativistic changes are caused by a system's motion through Mach Fields. If the existence of such fields is denied in

Special Relativity, then relatively moving systems cannot change each other. This is because a system cannot act across space to affect another.

Thus we have a theory which is more consistent than Special Relativity. For example, the asymmetric outcomes of time dilation experiments and the "twin clock effect" happen because clocks and particle decays experience asymmetric dilation during their motion. So, asymmetric results arise from previously asymmetric behaviour.

In the old way of thinking (Special Relativity), the Scientific Establishment (authors and scientists) are stuck with two major problems. Firstly, there is no actual evidence of the reciprocal dilation effect as predicted by Special Relativity, as all the results are asymmetric. Secondly, there is the problem of how to turn these reciprocal time dilations into an asymmetric result. The variety of explanations in books by scientists and authors (and the lack of unanimity) to explain the changeover demonstrates their implausibility.

The implications of Machian Relativity are truly profound. These are that not only high speed relativistic effects, but the more usual phenomena of classical physics that we experience on earth (such as the magnetism of an electric current, or the acceleration effects as a train slows down), are influenced by all matter in the universe, and this includes matter in the remotest of galaxies.

EPSILON III
Other Applications of Machian Relativity

The Theory of Machian Relativity allows a straightforward resolution of other paradoxes. For example, in the *Submarine Paradox*, the submarine has a much higher absolute velocity than the water it travels through. Therefore, it is physically shorter due to Lorentz contraction than when at rest. Accordingly, the submarine displaces less water and so sinks. In the *Pole-Barn Paradox*, the stationary pole is the same length as a barn, and just about fits inside it. But when accelerated to relativistic speed, its high absolute velocity makes it physically shorter, and so it fits inside the barn. Relative to the frame of the pole, the barn is longer, because it is essentially at absolute rest. And the same applies to the *TNT Paradox*. Let some explosive be placed in one end of a hollow cylinder. A plug fits over the other end, and projects inside the cylinder but does not quite touch the explosive (i.e. a plug-socket arrangement). If the plug is accelerated to high absolute velocity (relative to the external universe), it is physically contracted (while the cylinder is not), and so does not touch the explosive when it strikes the open end of the cylinder. However, if the cylinder with the explosive is instead accelerated to a high inertial velocity, it is now physically contracted instead of the plug. This means the plug has sufficient length to strike the explosive in the cylinder, causing it to explode. These explanations are far simpler than those of conventional science[3], which rely on the paradox-riddled Special Relativity that has to be combined with General Relativity or other phenomena to save it.

---

[3] Supposedly, the Twins Paradox "proves there is no absolute time", but this is bogus. The simplest reason for the asymmetric result is that time for the travelling twin is physically dilated for his whole journey, due to his relativistic velocity relative to the "Fixed Stars"; while (almost) nothing happens to the earth twin's time. So there is a *mixture of times*; the earth twin's Absolute Time, and the travelling twin's Relative Time. It is *a false premise* to claim (without considering alternatives) that because one observer experiences relative time, then other observers must be the same, so that inertial clocks change reciprocally relative to each other. Claiming that there can be no absolute motion because the speed of light is constant for inertial frames is a similar *fallacy*. And it is *rhetoric* to claim that "the equations of Special Relativity are kinematical, not physical". Einstein's interpretation of the Michelson-Morley null result (no Aether or any other absolute motion) is *confirmation bias*, and is contradicted by stellar aberration and the dilation of earth clocks in the Hafele-Keating Experiment. There is no evidence for specifically reciprocal dilation and it can be shown that covariance is an apparent effect due to physical changes. Acceleration effects may be absent during inertial motion, but the external universe is never absent, and neither are its effects. Its gravitational fields are always present, and so are its other effects.

ZETA

THE RELATIVITY OF NON-INERTIAL SYSTEMS AND MACH'S PRINCIPLE

Applying the Equations to Non-inertial Motion

The equations of Machian Relativity apply equally well to acceleration, because the equations for inertial motion are already asymmetric. This applies to motion along any linear, curved or polygonal line. Consider time dilation:

$$T_B = \gamma T_\alpha \qquad T_\alpha = T_B / \gamma$$

Where; T_α is the duration measured by a system at rest relative to the external universe and T_B is the corresponding duration measured by a system with absolute motion relative to the external universe. For both inertial and circular motion, the constant absolute speed v_α is used. If the speed is not constant, the instantaneous speed is used to calculate instantaneous time dilation. If an inertially moving clock passes the origin of a system at rest relative to the Fixed Stars, and clock readings are initially zero, the relative rate of the moving clock is:

$$t_B / t_\alpha = \gamma = \sqrt{(1 - v_\alpha^2/c^2)}$$

Which is constant as v_α is constant. If instead the clock is initially at rest relative to the origin, and undergoes constant linear acceleration a_α relative to the Fixed Stars, we have $v_\alpha = a_\alpha t_\alpha$. Hence the instantaneous relative rate continuously changes, and is given by:

$$\partial t_B / \partial t_\alpha = \gamma(t_\alpha) = \sqrt{(1 - v_\alpha^2/c^2)} = \sqrt{(1 - a_\alpha^2 t_\alpha^2/c^2)}$$

The average velocity is $\langle v_\alpha \rangle = (v_\alpha 1 + v_\alpha 2) / 2$, and for small durations, $t_\alpha 1 \approx t_\alpha 2$, the variation in t_B can be assumed as approximately linear. Thus the average dilation is:

$$\langle T_B \rangle \approx (t_\alpha 2 - t_\alpha 1) \sqrt{(1 - \langle v_\alpha \rangle^2/c^2)}$$

This approximation method can also be used for any arbitrary acceleration. The actual average dilation is found by integrating $t_B = t_\alpha \sqrt{(1 - a_\alpha^2 t_\alpha^2/c^2)}$ over the duration $(t_\alpha 2 - t_\alpha 1)$, then dividing by the duration. For example, the previous equation for constant acceleration gives:

$$\bar{T}_B = \left[\frac{-c^2}{3a_\alpha^2} \left(1 - \frac{a_\alpha^2 t_\alpha^2}{c^2} \right)^{\frac{3}{2}} \right]_{t_\alpha 1}^{t_\alpha 2} \Big/ (t_\alpha 2 - t_\alpha 1)$$

Thus, for acceleration by applied forces, in space free of the gravitational fields of massive bodies, there is no need to use General Relativity and its "induced gravitational fields". The cause of the asymmetry (like for inertial motion) is the Plenum Π of Mach Fields of the External Universe (Figs 27 and 28):

⊕ = CLOCK WITH ABSOLUTE MOTION

O = CLOCK AT ABSOLUTE REST

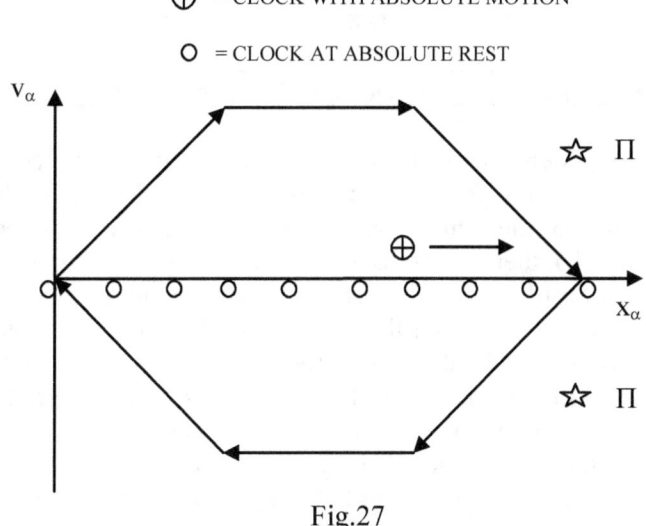

Fig.27

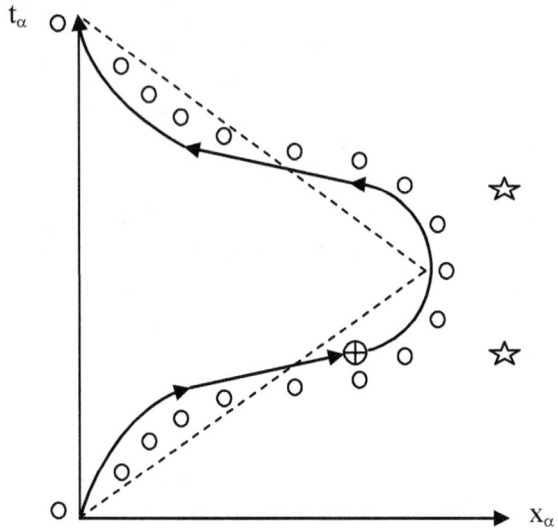

Fig.28: The Space-time Diagram for Fig.27

This is consistent with the current asymmetric use by conventional science of Special Relativity. Thus, General Relativity only becomes relevant in the presence of actual bodies with strong gravitational fields.

ZETA I
LINEAR ACCELERATION:
The Acceleration of Two Separate Clocks by identical forces

Consider two identical clocks at rest in the Plenum, separated by a distance L. If both clocks are simultaneously accelerated, each by applied forces of the same magnitude and direction (e.g. by rocket thrusts along L), then is this distance Lorentz contracted relative to observers at absolute rest? If each clock has a certain acceleration when in isolation (far from the other), then it would be expected that they must accelerate in exactly the same way when their separation is L. That is, each clock continues to obey Newton's Second Law of Motion. This is the simplest explanation. Therefore, they maintain their constant separation L during acceleration and also during inertial motion after the accelerations cease. So I say the clocks *do not* get closer due to Lorentz contraction. Relative to an observer moving with the clocks, the distance between the clocks (apparently) increases according to L/γ. This occurs because of the Lorentz contraction of his measuring rod.

However, if the space between both clocks *did* undergo Lorentz contraction, then only the clock defined as the origin would accelerate according to Newton's Second Law, while the other clock would not. Also, the choice of which clock is defined as the origin is arbitrary, so this behaviour would be reversed if the other clock was chosen as the origin. Now, we do not expect the behaviour of bodies to vary according to our arbitrary choices. Thus, Lorentz contraction *only occurs to solid matter, but not the space between objects.* So, the distance between the clocks can only be length contracted if they are attached to either end of a rod.

Corollary

Now consider the same two clocks as before. Both clocks are at absolute rest and synchronised. If both clocks are subjected to the same simultaneous forces, the simplest explanation is that they behave the same as each other in all ways, including their time dilations. Therefore, relative to observers at rest in the Plenum, both clocks are dilated in exactly the same way during their acceleration. Therefore, relative to these observers *the mutual synchronisation of both clocks with each other is maintained.* And this synchronisation is preserved during inertial motion after the acceleration, even though the readings of the clocks are delayed relative to clocks at absolute rest. Therefore, the two clocks do not have the relativistic synchronisation as required by the simultaneity equations of relativity. The relativistic synchronisation is achieved by adjusting the readings on either (or both) of the clocks accordingly. This adjustment can be done by observers at rest or moving with the clocks. Once this has been done, the speed of a light beam moving from one moving clock to the other is constant. On this basis, it must be concluded that relativistic simultaneity is, unlike time dilation, *not* an inherent property of time.

However, if relativistic simultaneity *was* an inherent property of time, then the readings on the two moving clocks would change accordingly during acceleration to give the required differences in readings. Although this now matches the equations, it would cause a variety of problems. Firstly, it would be expected that during acceleration, a clock's time dilation at any instant would be determined by its absolute velocity at that instant. But this would not happen if the clock's readings instead have to change to match the changes required by relativistic simultaneity. Secondly, there is the problem of deciding which clock is the origin of the system. If the first clock is defined as the origin, then it undergoes time dilation during acceleration, while the second clock will change differently to allow for relativistic simultaneity. However, this outcome is reversed if the second clock is instead defined as the origin. This occurs because the choice of the system origin is entirely arbitrary. No, both clocks must change in the same way as each other during acceleration as described earlier.

Corollary

During acceleration, the time dilation of most clocks will be determined by their instantaneous absolute velocity. This includes atomic clocks, radioactive decay of particles and biochemical processes (such as ageing). However, the functioning of mechanical clocks may be affected by the forces during acceleration (e.g. if the g-forces are much greater than earth's gravity). This will occur in combination with the usual time dilation. Unless the clock has been permanently distorted, this mechanical effect will disappear when the applied force is released. Mechanical clocks may also be affected by gravitational free-fall, but this will be due to a lack of g-force effects.

Accelerated Rigid Rod

Consider a rigid rod of length L at rest relative to the external universe. If it is accelerated by applied forces, it undergoes Lorentz contraction, and its instantaneous length is determined by its instantaneous absolute velocity relative to the external universe. The issue of which end (or other part) of the rod acts as the origin

Glyn Phillips

cannot now be avoided, because both ends are physically connected by the rod's matter. This contraction must occur around the rod's centre of mass, so that each end of the rod moves with a combination of its acceleration and its contraction to the centre of mass. If others assert that the contraction occurs at one end, then the behaviour of the rest of the rod during acceleration is dependent on the arbitrary choice of origin, which must be regarded as unlikely.

Corollary

Let there be two identical rods at absolute rest, placed end to end. If they are accelerated by identical applied forces, then both rods are Lorentz contracted relative to their respective centres of mass. Therefore, a gap will open up between the rods. This is because they are not physically attached. And like the previous case for accelerated clocks, each rod continues to behave as a separate body according to Newton's Second Law (F = ma).

Accelerated Non-rigid Rod

If a rod is non-rigid, then its elastic properties are relevant during acceleration. If a rod is accelerated from behind by applied forces, then it will be compressed during acceleration. Relative to stationary observers, the rod will initially have a low absolute velocity. Thus Lorentz contraction will be insignificant, and the rod's length will be determined by the mechanical compression of the applied force. As the rod's velocity becomes relativistic, then the compressed rod undergoes Lorentz contraction. So, stationary observers determine the rod's length to be shortened by a combination of mechanical compression and Lorentz contraction. When the applied force is released, the mechanical compression vanishes. The rod them coasts inertially with its length now only shortened by Lorentz contraction. If the rod is accelerated from the front end, the behaviour is similar, but the rod is instead mechanically stretched.

79

ZETA II
The Relativity of Rotating Systems according to Mach's Principle

Time Dilation in Rotating Systems

Consider a rotating system comprised of rigid radial spokes. A clock is placed at the end of each spoke. This creates a ring of rotating clocks. According to Special Relativity, the clocks are time dilated relative to stationary clocks. This change is presumably asymmetric, because the rotating clocks are non-inertial. But the asymmetry is merely assumed, and (apart from vague references to General Relativity) is never derived from first principles.

In the Theory of Machian Relativity, the asymmetry is derived from first principles. Thus, each rotating clock (a) is asymmetrically time dilated relative to stationary ones (b), due to its absolute motion relative to the external universe (Π). This dilation is maintained if the clock is released from the rotating system and moves inertially (Fig.29):

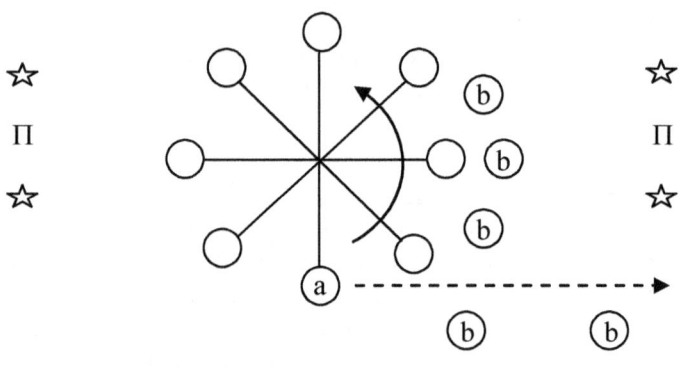

Fig.29

In General Relativity, a clock at a higher gravitational potential works faster than one at a lower gravitational potential. This has been proved in various experiments with atomic clocks. And this is a proper asymmetry - a clock at a lower gravitational potential works slower than one which is higher. The phenomenon of

Glyn Phillips

asymmetric time dilation in Machian Relativity is more in keeping with this asymmetric gravitational time dilation than the symmetric dilation predicted by Einstein's Special Theory.

Clock Simultaneity in Rotating Systems

As for relativistic simultaneity, Special Relativity does not give an adequate answer. If one clock on the ring is (arbitrarily) defined as the origin, then the simultaneity of successive clocks changes further around the ring. This creates a paradox when going full circle back to the original clock, because the theory predicts that its simultaneity should be both changed and unchanged. Conventional science then says that General Relativity is required for a proper analysis because the rotating system is non-inertial, but no such analysis is ever given.

This paradox is resolved by Machian Relativity, where relativistic simultaneity is not an inherent property of time. During the application of a force (turning moment) that causes the system to begin rotating, the clocks undergo time dilation according to their instantaneous absolute velocity. This is the same process as described earlier for linear acceleration. This continues during rotation after the applied force has ceased. Thus relative to observers at rest, the rotating ring of clocks is time dilated, but without relativistic simultaneity. The clocks can be resynchronised to produce relativistic simultaneity. This requires a reference clock to be arbitrarily defined, and creates a discontinuity, but this is no longer a problem as the simultaneity is no longer inherent to time per-se. Once this has been done, the speed of light is (locally at least) constant for observers rotating with the clocks.

Lorentz contraction in Rotating Systems

Consider a system of radial spokes attached to a central axis. The rods (spokes) of the rotating system are unchanged relative to stationary observers. This is because they are assumed to be an idealised rigid material, meaning centrifugal effects are assumed to be insignificant. Also, the motion of the rods is transverse to the external universe, so they are not Lorentz contracted along their radial lengths. If the spokes are non-rigid then they will be extended by centrifugal effects, without Lorentz contraction.

A rotating rigid ring (with no central spokes or other material) will undergo Lorentz contraction, and so its radius will physically decrease relative to stationary observers. This occurs because the ring approximates to a polygon with sides of infinitesimal length, each of which is Lorentz-contracted (like an inertially moving rod). Observers moving with the ring detect no change in the ring circumference, because their rulers are similarly contracted. However, they measure the radius to be *less* than when at rest - firstly, because their rulers are *not* contracted when placed radially – and secondly, because the radius of the ring has physically decreased. Thus stationary observers determine the circumference to always be $U = 2\pi R$ (for decreasing values of R), while rotating observers measure $U' > 2\pi R'$. In contrast to my reasoning, Einstein considers a ring of small rods placed end to end. This gives $R' = R$, and $U' > U$ (the ruler used by rotating observers is contracted only when placed along the circumference), so that $U' > 2\pi R'$ (the same as for a solid ring). Gaps must open between the rods on the circumference, because their ends are not physically attached. So this arrangement does not have a contracted radius for observers in either system. In addition, his conclusions are based on the predictions of Special Relativity, which has inertially moving systems and reciprocal Lorentz contractions. So, it is not explained how these asymmetric results for a rotating system arise. Admittedly the rotating system is non-inertial, but the asymmetries for dilation and contraction are merely assumed - they are not derived from first principles or described by properly asymmetric equations (there are usually references to General Relativity, but Machian Relativity properly explains such asymmetries in terms of Mach's Principle).

According to Einstein, Euclidean geometry does not apply to the rotating system. Instead, he deduces the rotating system is "at rest" in a gravitational field, with a non-Euclidean geometry (curved space-time). He says that the general laws of nature hold good for any change of coordinate system, and the equations are "generally covariant". However, I have a far simpler explanation for the measurements made by rotating observers. According to the Theory of Machian Relativity, the rulers in the rotating system undergo a Lorentz contraction which is a proper physical change relative to the external universe, and is objectively true for both stationary and rotating observers. Therefore, rotating observers conclude that the violation of Euclidean geometry, $U' > 2\pi R'$, is *apparent*, and arises from the contraction of their own rulers. They can verify this by directly comparing their own rulers to neighbouring stationary ones. Therefore, I say <u>the true law of nature</u> is that all systems change due to motion relative to the "Fixed Stars", rather than Einstein's method where all systems are somehow "at rest". There is nothing unusual about factoring out apparent effects. Consider for example, the retrograde motion of Mars relative to the earth. Since ancient times it was thought that such retrograde motion about a stationary earth was physically real. Since Copernicus, we now know that such motion is *apparent*, and arises due to the different orbital motions of the earth and other planets around the sun.

For a non-rigid ring, its radius will increase due to centrifugal effects, but this will be modified by Lorentz contraction. As for a rigid solid disk, this can be approximated to an outer ring surrounding a smaller inner disk. As the system rotates, the outer ring's tendency to Lorentz contraction is prevented by the rigid inner section. This puts the outer section under mechanical tension. If the disk is non-rigid, then the outer section's tendency to contract will compress the inner section. However, because the system is non-rigid, the whole system also experiences the additional effect of centrifugal inertia.

ETA

EXPERIMENTAL EVIDENCE AND RELATIVITY

Special Relativity is *self-evidently false*. The symmetric aspect has never been demonstrated experimentally, and there is no practical application of it in the scientific literature. All existing experimental evidence is asymmetric in nature, and is better explained by Machian Relativity.

Atomic Clocks

The results of the Hafele-Keating Experiment show that clock rates on aircraft and those on earth change asymmetrically relative to a non-rotating frame moving with the earth in its orbit. There is *no reciprocal dilation effect* between the aircraft and earth clock, other than an asymmetric change in their relative rates. The dilation of these clocks is asymmetric and relative to the average momentum of the Universe. Only the dilations of these clocks due to their different rotational motions (relative to the earth's centre) are relevant to the calculation. The clocks also have an additional dilation due to the earth's orbit of the sun, but this effect can be factored out because it is the same for each clock.

The Michelson-Morley Experiment

Because the earth's rotation has to be included in the Hafele-Keating Experiment, this *flatly contradicts* Einstein's idea that the Michelson-Morley null result is because there is no absolute motion and that observers on earth can regard themselves "at rest". On the contrary, because the earth's absolute motion influences the time dilation of its clocks, it must also induce a Lorentz contraction in the Michelson-Morley apparatus. This counters any "Machian wind" effect, giving the null result.

Relativistic Muons

The time dilation of muons is also due to absolute motion relative to the external universe. This includes muons produced in the upper atmosphere by cosmic rays, and muons in particle accelerators. The absolute velocity of such particles is near the speed of light and is much greater than the earth's absolute velocity. Thus, the earth approximates to a state of rest in this example.

Conclusion

Einstein's idea that there is no absolute motion and observers on earth can regard themselves "at rest" is *patently false* for a variety of reasons. We know from *mechanics* that the earth's daily rotation generates Coriolis effects, and rockets are launched in the direction of the earth's rotation to maximise their altitude. And the laws of gravity and celestial mechanics tell us that the earth is orbiting the sun. And in the Hafele-Keating Experiment, the earth's rotation affects the time dilation of clocks on earth.

We also know that the earth's annual orbit results in stellar parallax and aberration effects, together with Doppler asymmetries in the Cosmic Microwave Background Radiation. So the motion of starlight is relative to a frame independent of the earth's motion, and these phenomena constitute a form of "Aether wind" effect which Einstein claimed to have refuted.

And from De Sitter's Principle, we also know that the motion of light is independent of its source velocity. This means that light from any source in the universe moves relative to this frame, independently of the earth's motion. This must include light sources on earth, such as in the Michelson-Morley Apparatus. This is Mach's Stellar Parallax Principle. This absolute frame for the motion of light must be the same frame that generates non-inertial effects, time dilation and Lorentz contraction. According to Machian Relativity, this frame is the average momentum of the Universe.

So, the *motion* of all light is relative to the average momentum of the universe, causing "Aether wind" effects (stellar parallax and aberration) for observers on earth. But despite this, the *speed* of light for observers on earth is constant (the Michelson-Morley null result) due to Lorentz contraction, which counters any "Aether wind" effects. Because the earth's orbit of the sun is inertial, these effects must persist even if the earth or any other frame moves uniformly in a straight line according to Newton's First Law.

The concept of absolute motion and asymmetry is also inherent in the derivation of Time Dilation using Light Clocks. The change in the moving system's clock reading (from the classical value) is necessary for its observers to measure the speed of light to have the same value as that measured by stationary observers. So this change must be objectively true for both sets of observers, and is not just "how a moving clock looks relative to an observer". The cause of the asymmetry is the action of the Universe on the Light Clock. The situation is therefore not reversible.

Summary of the Hafele-Keating Experiment

- The time dilations of earth and airplane clocks *are not reciprocal*. They are asymmetric relative to non-rotating frames.
- Clocks on earth are dilated by the earth's rotational motion, which *refutes* Einstein's idea that the Michelson-Morley null result is due to the earth being "at rest". Thus the null result is due to the apparatus undergoing physical Lorentz contraction due to the earth's motion, which counters the "Machian wind" effect, keeping the speed of light constant for earth observers.
- Like clocks on earth and aircraft, an uncorrected clock (not corrected to counter Special Relativity) orbiting the earth must also undergo physical time dilation. Because such a clock is locally inertial, such physical dilation must also apply to a clock in *uniform rectilinear motion* relative to the earth. This refutes both the symmetry of Special Relativity, and the idea that being "non-inertial" causes the asymmetry. This also gives the simplest possible explanation for the asymmetric ageing effect in the Twins Paradox.

Glyn Phillips

A Brief History of the Michelson-Morley Experiment

In 1887, the Michelson-Morley experiment failed to detect the earth's motion through the Aether. Then (according to most scientists and authors) nothing much apparently happened until 1905, when Einstein working alone in a patent office had his "annus mirabilis" and developed Special Relativity by "pure thought". This supposedly not only disproved the existence of the Aether, but of absolute motion per-se, thereby "explaining" the null result. But that's not the complete story. Michelson believed the null result was due to the earth dragging the Aether, and that performing the experiment further from the earth (on a mountain) might yield a positive result. Fitzgerald instead developed relativistic length contraction in a stationary Aether, which compensated for the "Aether wind" effect. Lorentz then developed his Aether Relativity in 1904, which predicted both length contraction and time dilation as physical (asymmetric) changes due to absolute motion relative to Maxwell's stationary Aether. But this book proves that while the Aether does not exist, neither does Einstein's concept of purely relative motion – no experiment has ever shown dilation to be reciprocal. Other phenomena such as stellar aberration, parallax and asymmetries in the Cosmic Microwave Radiation clearly show the earth's motion is absolute. That is, there is a Universal Reference Frame for light and the earth is moving relative to it. And the Hafele-Keating experiment shows that the earth's rotation causes physical time dilation of clocks on its surface – this is a "dilation wind" effect that *is not a "null result"*. Thus the Michelson-Morley apparatus must be *correspondingly physically contracted*. The null result proves Speed of Light Constancy – it does not disprove absolute motion[4]. The assumption by Einstein and others that the null result refutes the existence of absolute motion (when there is an alternative explanation and contradictory phenomena) is an example of *confirmation bias*.

---

[4] Einstein's reasoning also contains the following epistemological defect; if the null result is due to the earth being "at rest", rather than the apparatus being physically contracted, then the Lorentz contraction predicted by his equations serves no physical purpose, other than as a form of optical illusion for observers in relative motion.

THETA

TESTING FOR RECIPROCAL TIME DILATION

Einstein's Theory of Special Relativity is based on two postulates. Firstly, that the Laws of Physics are "the same" for different inertial systems. Secondly, that the speed of light as measured by relatively moving inertial systems is constant, regardless of the relative motion between the system and the light source.

The problem with this theory is in the first postulate, known as the Relativity Postulate. According to conventional science, this postulate requires complete symmetry between inertial systems, because supposedly, absolute motion and absolute rest do not exist. This idea came about in two ways. Firstly, the standard reasoning used to justify this is the Michelson-Morley null result. According to Einstein, because no "Aether wind" effect could be detected, then the Luminiferous Aether (or any other state of absolute rest) could not exist. Secondly, in deriving the space-time equations from the Lorentz Transforms, Einstein assumed that their covariant form (reciprocal symmetry) mathematically disproved absolute motion, and no other interpretation for it was considered. So, the Michelson-Morley null result became "experimental proof" of the Relativity Postulate, while the covariance of Special Relativity became "theoretical proof".

But, because the equations for time dilation are covariant, this leads to the Clock Paradox problem, which conventional science claims to have resolved. Obviously, the paradox doesn't happen in real life - from experiments on atomic clocks and planes, we know that the travelling twin will be younger than his twin when he returns. This is not the paradox. The paradox is in Einstein's theory and the various "explanations" concocted by physicists that try to get the required asymmetric result from its symmetric equations.

Unfortunately, physicists haven't made up their minds on which is the proper "explanation" or experimentally verified it.

Now, books written by scientists and authors claim that Special Relativity "has been proved in time dilation experiments". However, this claim is completely false. Firstly, the experiments usually involve systems moving non-uniformly; and secondly, all the results are asymmetric and consistent with the systems undergoing a proper physical and asymmetric change. So such experiments do not specifically demonstrate the reciprocal aspect.

Consider for example, the Hafele-Keating Experiment, where an atomic clock is flown around the earth on an aircraft and its reading is compared to a similar clock on earth. Assume that the effects of General Relativity (altitude) on the aircraft clock rate have been allowed for. In order to calculate the rate of the aircraft clock relative to the earth clock, the respective time dilations for both clocks due to their motions relative to a reference clock at the centre of the earth are calculated. In other words, the time dilations for both clocks are assumed to be normal asymmetric and physical changes relative to another frame of reference. So, there is no reciprocal effect between both clocks, and their time dilations are not determined by their relative motion. Also, scientific papers describing the experiment make no use of the reciprocal aspect of Special Relativity.

Therefore, Einstein's Relativity Postulate has not been specifically tested. This can only be done by conducting tests on relatively moving inertial systems, which factor out any acceleration effects. For scientists and authors to claim that relatively moving clocks work slower relative to each other *is an extraordinary claim*, and extraordinary claims require extraordinary levels of proof. Because of this problem, I have devised three possible experiments to resolve the issue.

Experimentum Crucis I

Consider the Hafele-Keating Experiment, but apply it instead to an atomic clock (a) orbiting in space around the earth (Fig.30). Such a clock might be on a GPS satellite or a spacecraft. Let the clock be pre-adjusted so that the effects of General Relativity are factored out. This means the clock will only show the effects of Special Relativity when in orbit. Let the clock orbit in the direction of the earth's daily rotation (west to east). Every time the GPS satellite (a) passes over the earth clock (b), its clock reading is sent to earth as a digitised signal (downlink, d). Observers on earth can then compare their clock readings with those of the satellite.

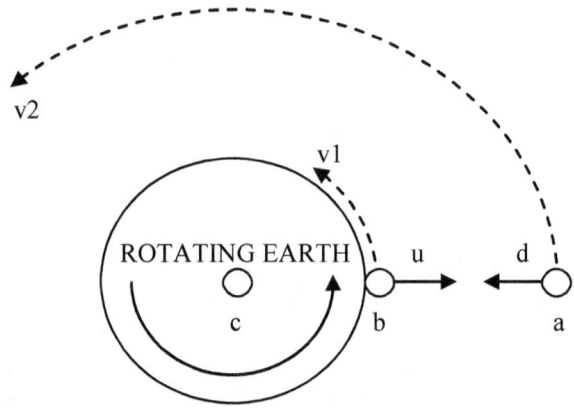

Fig.30: Modified Hafele-Keating Experiment

Like the Hafele-Keating experiment with aircraft clocks, the experimental results will show that the satellite clock has a relatively slower rate during its orbit (likewise, the satellite clock rate will be relatively faster if instead it orbits east to west against the earth's rotation). The theoretical relative rate predicted by Special Relativity is calculated from the respective time dilations of the earth and satellite clock relative to a third reference clock. This is done using the rotational velocities of the satellite and earth clock relative to a clock in a non-rotating reference frame, which is provisionally placed at the centre of the earth (c). In other words, the satellite clock has a <u>partial</u> time dilation relative to the earth clock. It is known the theoretical predictions and experimental results are in good agreement.

Now, papers on the Hafele-Keating Experiment (where the moving clock is on an aircraft) never explicitly state that the time dilations are asymmetric, although from the way the relative clock rate is calculated, this must be the case. Certainly, there is no mention or use of the reciprocal aspect of Special Relativity. The reference clock is assumed to be inertial because it is non-rotating, while the earth and aircraft clocks are non-inertial because they are in rotating frames. The rotating clocks are then assumed to change physically relative to the non-rotating clock because non-inertial frames "feel force effects".

Also, there is no mention of what the corresponding rate of the earth clock relative to an observer moving with the satellite or spacecraft might be, or what equation might describe it. Because the time dilations are physical changes, the earth clock must have a faster relative rate to the satellite clock.

Relative to the earth and satellite clock, the reference clock will have a particular <u>inverse dilation</u> (faster rate). Each inverse-dilation value is calculated from a normal algebraic rearrangement of the corresponding time dilation equation. Thus relative to the satellite clock, the earth clock will have a <u>partial</u> inverse dilation.

Therefore, the experiment can be modified to measure this - in addition to the downlinks to the earth clock, readings on the earth clock can be sent to the satellite clock (uplink, u). This will allow a direct measurement of the rate of the earth clock relative to an orbiting observer. The value can also be calculated theoretically from the method of time dilations of the satellite and earth clocks relative to the earth-centred clock.

Such an experimental modification will confirm that the earth clock has a correspondingly faster rate (partial inverse dilation) relative to the orbiting satellite clock (this is consistent with the faster relative rate of aircraft clocks relative to earth observers when they are flown counter to the earth's rotation). This means the following:

i)   There is no reciprocal dilation effect between the earth and satellite clock. The clocks are not working slower than each other due to their relative motion.

ii)   Both the earth clock and the satellite clock are undergoing asymmetric time dilation relative to the earth-centred reference clock.

Points (i) and (ii) have the following profound implications:

Firstly - This experiment has not demonstrated the reciprocal aspect of Special Relativity.

Secondly - Because the motion of the earth clock has to be considered in the calculation, observers on earth cannot consider themselves "at rest". Therefore, Einstein's previous interpretation of the null result of the Michelson-Morley Experiment as being due to earth observers being "at rest" because "there is no absolute motion" must be false.

Thirdly - The orbiting satellite clock is experiencing asymmetric time dilation, yet (unlike the aircraft clock) it is not experiencing force effects because it is in gravitational free-fall. In fact, according to Einstein's Equivalence Principle, frames in gravitational free-fall are equivalent to inertial frames. So, if such a clock is inertial and has asymmetric dilation, it is likely *a similar clock moving inertially from the earth in a straight line* will experience the same asymmetric effect. So Einstein's hypothesis of reciprocal dilation for relatively moving inertial frames must be false. It also means that the conventional belief that "acceleration effects cause asymmetry" (such as in the Twins Paradox) must also be false.

This allows for a very straightforward explanation of the Twins Paradox. If a clock (or twin) goes on a relativistic journey from the earth and returns, its time is physically dilated (slowed) for every part of its journey. This includes during its inertial phases. However, time for the earth clock (or twin) will not be dilated. Thus relative to the travelling clock (or twin), the earth clock (or

Glyn Phillips

twin) will be correspondingly advanced. This explains the asymmetry in times (or ageing) on reunion.

Also, while the orbiting clock experiences no force effects, the earth clock does, because it is not in free-fall. Therefore, if force effects are necessary to make a frame change physically, it would be expected that the earth clock would instead be physically time dilated relative to the satellite clock, but this does not happen. Clearly, physical time dilation is happening due to motion alone, independently of force effects, so the conventional hypothesis for explaining the asymmetry must be false. In addition, imagine two atomic clocks orbiting the earth in opposite directions. In this case, both will be inertial, yet they will be equally dilated relative to the earth-centred reference clock. Thus, they experience no reciprocal dilation despite being in relative motion and inertial.

Experimentum Crucis II

Consider two people on earth – one is an astronomer and the other is an astronaut. Both have telescopes and lasers, which emit the same pulses at regular intervals. The astronaut accelerates from the earth and then moves at constant relativistic speed (Fig.31). Each person then observes the other through his telescope:

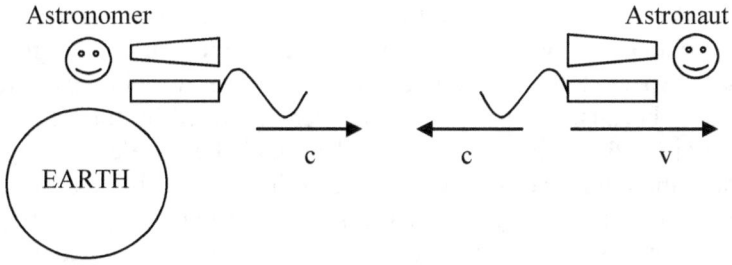

Fig.31

If the Relativity Postulate of Special Relativity is *correct*, then there will be complete symmetry in what each person observes during the inertial motion. Using their telescopes, each person will

93

see the other slowed due to a combination of time delays (due to their increasing separation) and time dilation. The same applies to the laser pulses – each person will see the other's laser pulse intervals to increase due to a combination of time delays and time dilation.

However, if the Relativity Postulate is *incorrect*, then time dilation will be asymmetric during inertial motion, and there will be a corresponding asymmetry in what each person observes. Using his telescope, the astronomer will see the astronaut slowed due to a combination of time delays and time dilation. However, the astronaut will see the astronomer partly slowed due to time delays and partly quickened due to inverse dilation. The same applies to the laser pulses. The astronomer will see the astronaut's pulse intervals increased due to a combination of time delays and time dilation. But the astronaut will see the astronomer's pulse intervals partly increased due to time delays, and partly reduced due to inverse dilation, giving an average effect[5].

Experimentum Crucis III

Consider the usual experimental setup. Radioactive subatomic particles (such as muons) have been accelerated to near the speed of light in a circular accelerator, and are now moving at a constant speed (Fig.32). After a certain time, the number of remaining particles can be counted by diverting them to a detector (a). Experiments show that the number of remaining particles is greater than expected (e.g. in comparison to slower moving particles). This is because their half-life has been extended by time dilation. Similarly, detectors can be placed around the accelerator (b), to count the number of decay products emitted by the particles (while they are still moving). This gives a lower than expected result, which means their decay rate has been reduced. This is consistent with the extended half-life.

---

[5] If the astronaut returns, Machian Relativity predicts the astronomer always has inverse dilation relative to the astronaut (apparent effects cancel on reunion). The General Relativity "explanation" predicts a compensatory faster rate (relative to the astronaut) during the turnaround phase, while the Time Delays "explanation" predicts it during the return.

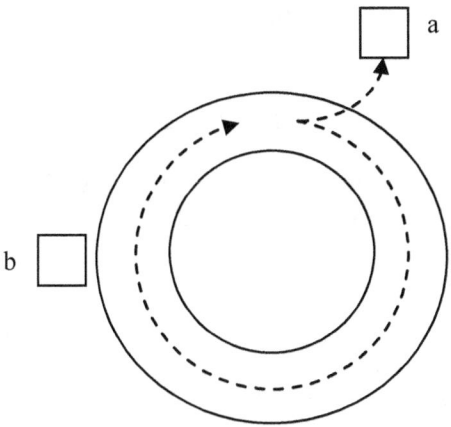

Fig.32: Muons in an Accelerator

We know this change is properly asymmetric and physical, because there is a <u>reduction in the transfer of matter</u> from the high speed particles (while they are still moving) to the detectors which are at rest on the earth. So, this process cannot be dismissed as being "kinematic" or "how something looks relative to an observer". Thus, relative to the moving particles, there must be a corresponding <u>increase</u> in the number of decays from unaccelerated particles on the earth (this is consistent with the faster working of atomic clocks when they are flown counter to the earth's daily rotation). However, such faster working of earth clocks is apparent, because the true cause arises from the greater physical dilation within the faster moving system itself.

This asymmetry in results is explained by the Theory of Machian Relativity, and arises due to absolute motion relative to the average momentum of all matter. Scientists and authors do not provide an adequate explanation however. The asymmetry is either assumed (by ignoring one set of Einstein's covariant equations), or the issue may be evaded by references to General Relativity. Another problem with conventional science is that there is no explanation for what the asymmetry is relative to. For example, in the Twins Paradox, the asymmetry is assumed to be relative to the earth, but there is no reason why it cannot be relative to the sun or any other astronomical object. So the choice of asymmetry is completely arbitrary.

Testing Dilation during Inertial Motion

Now, to test the Relativity Postulate, this experiment can be modified (Fig.33) by introducing straight line sections into the circular accelerator (a-b, c-d):

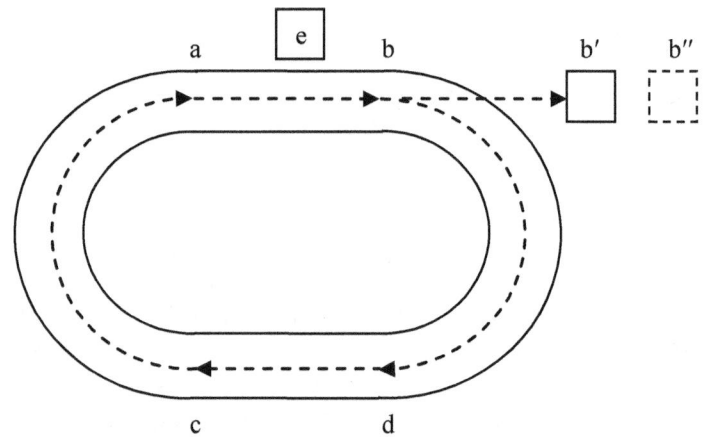

Fig.33: A Modified Accelerator

Particles are not accelerated along these sections, so that they coast inertially in straight lines. After a certain time, particles moving inertially along (a-b) are allowed to drift further to (b'). The number of remaining particles can be counted by placing a detector at (b'). Likewise, their decay rates along these sections can also be counted (e). If required, the experiment can also be repeated by increasing the length of the inertial motion by moving the counter from (b') to (b'').

Now, if these results are indistinguishable from the previous experiments were the particles moved non-uniformly (and it is a certainty they will be), then it must be deduced that these inertially moving particles are also undergoing asymmetric changes in their time dilation, in accordance with Mach's Principle. *Therefore,*

Einstein's Relativity Postulate has been refuted. If scientists and authors try to evade this conclusion, by saying that "symmetric and asymmetric results cannot be distinguished" or "the symmetry cannot be tested" then Special Relativity is not scientific, because its consequences cannot be tested.

Changing the Reference Frame for Time Dilation

It is clear that if the time dilation of an orbiting (and thus inertial) atomic clock is asymmetric as in the Hafele-Keating Experiment, then the usual explanation that non-inertial effects are causing the asymmetry is false, as there are no such effects. Something else must be causing the asymmetry. Now consider the earth orbiting the sun (Fig.34):

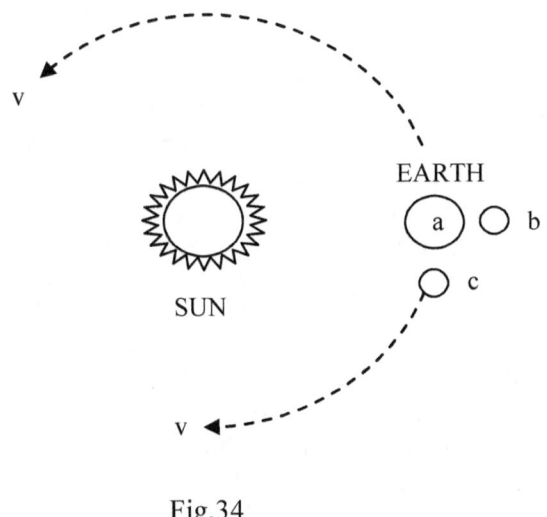

Fig.34

If the Hafele-Keating method is applied to this situation, then the earth-centred reference clock (a) of Hafele-Keating must also experience asymmetric time dilation relative to the frame of the solar system, depending on the earth's orbital speed (v) of the sun. So the initial Hafele-Keating assumption that this clock is "at rest" is false, especially considering there can be no reciprocal dilation of non-orbital "solar system" clocks relative to the earth. Clocks orbiting the earth also have this sun-orbital motion, so they must be time dilated by a combination of these two motions. But it is unsatisfactory that a body is at rest or moving (or time dilated or not) depending on our arbitrary choice of reference system. Rather, I say that bodies are time dilated by their absolute motion relative to the "Fixed Stars". Thus, because the earth and satellite clocks share in their motion of the sun (and galaxy), the effect of such motion affects both types of clocks equally, and so is factored out within the frame of the orbiting earth.

This analysis also gives some interesting deviations from the usual Twins Paradox scenario. Let a twin and clock accelerate from the earth and then remain at the same fixed point in the solar system (b). When the earth completes its orbit, the twin accelerates to return to earth. This means the twin on earth and the reference clock will be *younger* than the space twin. Thus the travelling twin experiences forces as in the Twins Paradox, but he is now the older one. Likewise, if the twin leaves the earth and goes in a gravitational retrograde orbit of the sun (c), when he returns both he and the earth twin will have experienced the same time dilation, so there will be *no relative difference* in their ages (ignoring effects from the earth's daily rotation). So clearly, the assumption that a moving twin or clock needs to experience force effects to make its dilation physical rather than reciprocal is <u>false</u>. Physical dilation happens in the absence in force effects. And the assumption that physical dilation happens relative to the earth, sun or any other celestial body only serves as a useful approximation. It is not what is actually happening. Likewise, there are instances where the travelling twin can return to earth and be older than the earth twin, rather than younger.

By applying the Hafele-Keating method to the frame of the earth or the solar system, it can be seen that time dilations are *always asymmetric*, even when clocks are in gravitational free-fall and thus locally inertial. Thus there can be *no objection* to a twin (or clock) also undergoing asymmetric time dilation when travelling inertially relative to the earth at high speed - especially considering that like a clock orbiting the earth, such a clock is also moving through the gravitational fields of the earth and <u>all other matter</u> in the Universe. Space is not a true void, but is a plenum of the gravitational fields of all matter.

An aircraft clock flying west to east experiences lower non-inertial effects than a clock on earth. Yet the aircraft clock has the greater dilation. Also an uncorrected clock (unlike a GPS clock) orbiting a *non-rotating earth* will be physically time dilated relative to the earth clock, even though the orbiting clock is inertial while the earth clock is non-inertial. Hence ascribing a clock's asymmetry to it being "non-inertial" <u>is bogus</u>. *Something else is causing it.*

The Clock Paradox Decoded

The *Simplest Explanation* for all these experimental results and the Twins Paradox is that the time dilation is physically asymmetric for all types of motion, inertial and accelerated. The reciprocal dilation of Special Relativity <u>DOES NOT EXIST</u>. Time dilation is *always asymmetric*, and the asymmetry is relative to a universal frame of absolute rest. I say this is the *average momentum of all matter*, according to **Mach's Principle**. This frame of reference approximates to the "Fixed Stars". Celestial bodies such as the earth or sun only approximate to a state of absolute rest, due to their low absolute velocity. The larger the mass of the body, the lower its absolute velocity is likely to be. As demonstrated elsewhere in this book, the Machian reference frame is self-generating from the relative effects of all matter, so there is no need to postulate additional phenomena that exist independently of matter, such as Maxwell's Luminiferous Aether.

This also *explains the Michelson-Morley Experiment* – the null result is due to a combination of a physical Lorentz contraction of the apparatus due to the earth's absolute motion <u>and</u> the absolute motion of light – both of which are relative to the "Fixed Stars". Einstein's interpretation of the null result as being due to the earth being "at rest because there is no absolute motion" is clearly false. We know from the Hafele-Keating Experiment that the earth's daily rotation causes a physical time dilation of clocks on the earth's surface. Thus the earth is NOT "at rest" for its observers. By extension, the earth's motion (daily rotation and annual orbit) must be causing a physical Lorentz contraction of the Michelson-Morley apparatus. This counters the "Aether wind" effect, thereby giving a null result, which keeps the speed of light constant for earth observers. Another inconsistency with Einstein's interpretation is that if the null result means there is no absolute motion, then the Lorentz contraction effect predicted in his equations serves no logical purpose. We also know that starlight has motion relative to a frame independent of the earth's motion. This is because of stellar parallax, aberration and Doppler asymmetries in the Cosmic Microwave Background Radiation - all of which occur due to the earth's annual orbit of the sun. So, these can be likened to an "Aether wind effect" or a "Machian wind effect". And this must also apply to the light source in the Michelson-Morley apparatus. However, conventional science makes no attempt to explain such phenomena in terms of Special Relativity.

The *covariance of the equations of Special Relativity* is an apparent effect only. If an inertial frame has absolute motion, the reciprocal time dilation it measures for a frame at absolute rest is an artefact of its own inherent relativistic simultaneity. When such simultaneity effects are factored out, the true measurements show that moving observers measure the rate of clocks at absolute rest to be faster (inverse dilation). This is how the paradox is resolved. The physical changes in frames with absolute motion maintain c-invariance for those frames. Frames at absolute rest do not need to undergo reciprocal change relative to moving frames, because light has absolute motion relative to the absolute frame – the Machian Plenum.

If the absolute motion of the earth is not significant, the travelling twin is physically dilated relative to the earth throughout his whole journey, including both inertial and accelerated motion. This is the simplest explanation, and avoids the flip-flopping of conventional physics, where symmetric dilation changes to physical dilation every time the motion changes from inertial to accelerated and vice versa.

The effects of *Time Delays* are <u>irrelevant</u> for the Twins Paradox analysis. This includes its effects on twins exchanging light signals, how they look relative to each other or Doppler effects. Such effects are apparent and (like perspective and parallax) have no physical consequences. The effects during recession are countered by effects during return, and completely cancel on reunion, leaving only the physical effects of his time dilation. Therefore such effects are completely peripheral to the Twins Paradox problem. The same reasoning applies to relativistic simultaneity. Such effects are not used to explain the time dilation of atomic clocks or relativistic particles, so there is no need to use them to "explain" the Twins Paradox. The physical dilation of the travelling twin can be directly measured by comparing his clock to synchronised clocks placed along his path, which are stationary relative to the earth.

As for *Minkowski Diagrams*, their use by scientists to "explain" the Twins Paradox is another irrelevance. Such diagrams are constructed around the symmetry of Special Relativity, and all they do is repeat the same arguments of Special Relativity, but in a graphical form rather than algebraically. Such *tautology* <u>does not</u> constitute a proof. Because the covariance is an apparent effect in Machian Relativity, such diagrams can be redrawn with the symmetry factored out, so that only the space-time axes of the frame with absolute motion are rotated. This is how they were originally formulated by Minkowski. The symmetric versions are later variations by other scientists. Their original purpose was purely illustrative.

Conclusion

According to *Popper's Principle* a scientific theory must be falsifiable. So, if scientists and authors persist in making claims that Special Relativity is correct, then they need to provide unambiguous experimental evidence for relatively moving clocks working slower than each other. But first they would need to admit that the current evidence only shows asymmetric dilation. The current vagueness on this issue in the scientific literature is certainly a mystery.

The evidence for absolute motion over relative motion is overwhelming. Consider electromagnetism – parallel streams of electric charges moving inertially relative to the earth are magnetically attracted. This must also be an objective fact for observers moving with the charges, thereby violating the Relativity Postulate. And observers on the earth see the stars to have an annual stellar parallax, proving the earth has a detectable absolute motion, due to its orbit of the sun. But if physicists try to deny that the symmetry can be tested, then Special Relativity is not a true scientific theory, because its consequences cannot be tested.

The same reasoning applies to a clock rotating about a fixed point. Its dilation is asymmetric relative to stationary clocks. This is an objective fact for observers at rest or moving with the clock. When released, it then moves away at a tangent, and according to Machian Relativity its dilation must remain asymmetric relative to stationary clocks. This is the simplest outcome. And according to *Occam's Principle* (Ockham's razor) if there are different interpretations, the simplest one is more likely to be true.

Clocks cannot work more slowly than each other because it is unphysical. And if the equations say it can happen, then the equations are <u>WRONG</u>. The equation $5 - 8 = -3$ is mathematically correct, but 8 oranges cannot be removed from a bowl of 5 oranges, because that is also unphysical.

Aristotle incorrectly assumed heavier bodies fell faster than lighter ones, and I say Einstein made a similar error – he incorrectly assumed how quantities should change relative to each other. Two inertial systems moving relatively to each other may feel the same for their observers, in terms of not experiencing the effects of applied forces, but that similarity cannot extend to a similarity of time dilation. And if one inertial system is relativistic, its observers see Doppler effects in surrounding stars which observers in the non-relativistic system do not.

IOTA

A REFUTATION OF ALL CONVENTIONAL "EXPLANATIONS" OF THE TWINS PARADOX

IOTA I

Firstly, let us begin with the <u>definitive solution</u> of the Twins Paradox. This will allow a more straightforward demolition of the usual explanations concocted by physics professors and science authors that readers may have encountered in other books or on the Internet. Consider the usual Twins Paradox scenario. One twin stays on the earth, while the other goes on a long journey at relativistic speed and returns to the earth. According to the theory of Machian Relativity, time for the travelling twin is physically dilated (slowed) **during his motion**, *relative to the Plenum* (average momentum of all matter), <u>not</u> due to motion relative to the earth twin. The earth twin has a lower absolute velocity, and so ages at a normal rate. This is because the earth and other celestial bodies, such as the "Fixed Stars", are difficult to accelerate to high speeds due to their high masses. So, the earth is a reasonable approximation to a state of absolute rest (in this scenario at least). Thus, the travelling twin is younger than the earth twin when they are reunited. ***This is by far the simplest explanation of the asymmetric ageing.*** That is, the travelling twin has aged continuously at an asymmetrically slower rate for <u>all parts of the journey</u> according to Mach's Principle (e.g. for both accelerated and inertial motion). NO OTHER SUPPLEMENTARY PROCESSES OR EXPLANATIONS ARE REQUIRED. And this is a proper physical change – relative to the space twin, time for the earth twin (in fact time throughout the whole universe) passes at a faster rate. But, this faster rate is *apparent*, and arises from the space twin's own dilated time. The external universe slows time for the space twin. But, the space twin cannot slow time in the external universe (nor physically cause it to go faster). This is because the external universe has more mass than the travelling twin, and in physics things with more mass tend to have a greater effect on less massive bodies than vice versa. Thus, Einstein's postulate of constant light speed (c) for inertial systems *is correct*, but the Postulate of Relativity (as he formulated it) is false. The average momentum of all matter in the universe defines a state of absolute rest, because momentum is always conserved according to Newton's Third Law. Conventional physics does not make use of this fact.

IOTA II

The Twins Paradox – Case Closed

In books and Internet forums, physics lecturers, authors and other supporters of Einstein argue that "the Twins Paradox is not a real paradox because the force effects experienced by the travelling twin cause an asymmetry", and "the equations of Special Relativity do not apply to his reference frame because it is non-inertial". But then they proceed to explain the asymmetry in terms of time delays due to increasing or decreasing distance between the twins, thereby making no specific use of force effects! But if this is true, then *under what circumstances* will the symmetric time dilation of Einstein's theory ever show itself? Could it be that the actual existence of such symmetry is as *bogus* as the Aether it supposedly refutes? More and more, the various "explanations" concocted by university physics professors, to get Special Relativity to give asymmetric answers, begin to look as implausible as Ptolemy's "wheels within wheels" method to explain planetary motions.

These explanations are also inconsistent in how they explain the asymmetry. Both rely on somehow generating a compensatory faster rate for the earth twin relative to the travelling twin during some part of the journey, to counter his usual dilation relative to the travelling twin. In the General Relativity version, this happens during the travelling twin's turnaround as he returns to earth. But in the Time Delays version, this happens during the travelling twin's return journey. No experiment has been proposed by scientists to resolve the issue.

The issue is whether these "explanations" (or *relativistic fudge factors*) genuinely describe the asymmetry as it occurs in nature, or whether they are "just so stories" concocted by university physics professors to save Einstein's theory from its illogical consequences. Once these problems are understood, Einstein's theory collapses like a house of cards.

PROPOSITION I: The Twins Paradox is not resolved by non-inertial effects

The claim that *"the asymmetry is because the space twin experiences force effects while the earth twin does not"* is patently false. Force effects *per-se* occur for only small parts of the journey, so they cannot possibly compensate for the symmetric effects that occur during the inertial phases – they can only generate asymmetric dilations while they are actually occurring. For example, if the inertial phases are varied while the accelerations are not, then the same accelerations would have to compensate for inertial phases of different durations, which is clearly ludicrous. Hence scientists' need for these other "explanations"!

Supposedly, *"the situation is asymmetric, because the travelling twin is non-inertial"*. But consider a journey which is 95% inertial, and 5% accelerated. What scientists and authors are trying to do is bury the 95% symmetric dilations in the 5% asymmetric dilations. A travelling twin cannot be regarded as being "non-inertial" if his motion is 95% inertial. And if the journey was 99.9% inertial and 0.1% accelerated, conventional science would still have us believe that the end result on reunion is 100% asymmetric. To expose this erroneous thinking, we can look at this in another way - if it's OK to bury the symmetry of 99.9% in the 0.1% asymmetry, then why not do the opposite, and bury the 0.1% accelerated (asymmetric) motion in the 99.9% inertial (symmetric) motion, to make the result for the twins on reunion 100% symmetric ? What is a paradox is scientists claiming the symmetry is a "merit", but then finding every imaginable excuse to make it vanish!!

PROPOSITION II: The Twins Paradox is not resolved by Time Delays

In the *Time Delays version*, decreasing time delays cause each twin to observe the other to have a faster rate during the return phase. But because the travelling twin observes this for a longer period than the earth twin does, this (somehow) causes the earth twin's total ageing to be advanced relative to the space twin on reunion.

According to this variation, both twins see the other slowed on the outward journey, by a combination of increasing time delays (a "Doppler Dilation") and reciprocal time dilation. While during the return journey, they see each partially slowed by reciprocal time dilation, and partially advanced by decreasing time delays.

But, another variation claims to directly derive asymmetric dilation for the travelling twin based solely on the effects of time delays on light signals reflected off the travelling twin back to the earth twin. This supposedly happens during the space twin's outward inertial motion from the earth, even before his turnaround has occurred.

It must be remembered that the inclusion of time delays into the analysis is <u>completely bogus</u>. Clock rates are determined by comparing relatively moving systems of clocks – this allows for a *direct comparison which factors out time delays effects*. For example, clocks can be placed in space along the path of the travelling twin or clock. These are at rest relative to the earth and synchronised with its clock. The effects of time delays on observed time dilations are apparent (e.g. like parallax). Thus, (as calculated elsewhere in this book) they <u>vanish on reunion,</u> so that only the true effect of the travelling twin's physical time dilation remains. The effects of time delays should be calculated from this physical time dilation, but not vice versa. To assume that time delays can have physically real effects is a <u>corruption of reasoning</u>. It is as ridiculous as saying *"because the returning twin sees the earth twin getting bigger due to parallax for a longer period than what the earth twin sees happening to the space twin, then the earth twin must be bigger than the travelling twin on reunion"*!!!

Glyn Phillips

Additional variations of this method also consider Doppler effects, where the twins see each other's light blue-shifted on the outward journey and red-shifted on the return. Supposedly, waves "cannot be lost" by the space twin, but they will be if his time is dilated.

Also, it may be claimed that the space twin registers less time for the journey because the journey's distance is Lorentz contracted in his frame. If this is true, then the earth twin will be correspondingly time dilated, and thus younger than the space on reunion. Also, "the travelling twin changes inertial frames, so is non-inertial". But each inertial frame should be the same as the other, as there is no absolute frame. To introduce an asymmetry, the motion of light is assumed to be absolute relative to the earth, so that the travelling twin's observations suitably change during the journey. But then it would be simpler if the dilation was also asymmetric during the journey.

PROPOSITION II: The Twins Paradox is not resolved by General Relativity

In the *General Relativity version*[6], the earth twin's compensatory faster rate relative to the travelling twin instead occurs during the travelling twin's turnaround phase only, because the associated accelerations supposedly generate a gravitational field (according to the Equivalence Principle) throughout space. Supposedly, the travelling twin is at rest in this field while the earth (and the rest of the universe) "falls back" to the travelling twin. Because the supposed gravitational field is directed toward the travelling twin, the earth is at a higher gravitational potential, which (according to General Relativity) means its clock rates are *faster relative to the space twin*. The initial and final "falling" of the earth (at the start and end of the journey) correspond to gravitational fields directed away from the travelling twin. These accelerations supposedly correspond to *lower gravitational potentials*, and thus *slower* earth clock rates relative to the space twin. However, because the

---

[6] This method doesn't consider what each observer sees due to Time Delays, although this is possible. However, their inclusion in the analysis is ultimately irrelevant, if it is recognised that their effects are apparent and are nullified on reunion (as in Machian Relativity).

turnaround phase occurs at a *much greater distance* from earth than the initial and final acceleration phases, the magnitude of its high gravitational potential (fast rate) cannot be compensated by the low potential phases (slow rate), which are insignificant. Thus the earth twin's clock has an overall advanced reading relative to the space twin on their reunion.

None of this is plausible however. As pointed out by Herbert Dingle, when the space twin returns, the synchronization of all clocks throughout the universe will be altered – all because of the motions of a single spaceship. Secondly, the induced gravitational field experienced by the space twin is ad hoc. Unlike the gravitational fields of celestial bodies, the induced gravitational field is not associated with a source mass, so the earth and other bodies cannot be mutually interacting according to Newton's Third Law. Also, the faster earth clock rate relative to the space twin during the turnaround phase means the space twin's correspondingly slower rate relative to the earth twin is instead determined by this induced field, rather than the space twin's relative motion per-se.

Another outcome is that if a second spaceship is simultaneously subjected to equal and opposite accelerations, it will perform an identical journey to the first spaceship, but in an opposite direction relative to the earth. Thus, its induced gravitational fields will be equal and opposite to the other induced fields, so their effects will mutually cancel. Hence, there can be no resultant field to explain how earth clocks are advanced when both spaceships return.

PROPOSITION IV: The Twins Paradox is not resolved by Space-time Diagrams

Conventional science also relies on space-time diagrams to further illustrate the asymmetry between the twins. It is said that the twins have different paths through space-time (Fig.35a). The earth twin (a) has a space-time path up the time axis, while the space twin (a') moves at an angle away from the time axis towards the space axis (outward journey). Then for the return journey, the space twin's path returns to the earth twin on the time axis. As the space twin moves through space (approaching the speed of light), his passage through time is reduced by time dilation (so that $t' < t$). But, the problem is, a similar space-time diagram can be drawn with the space-twin at rest (Fig.35b):

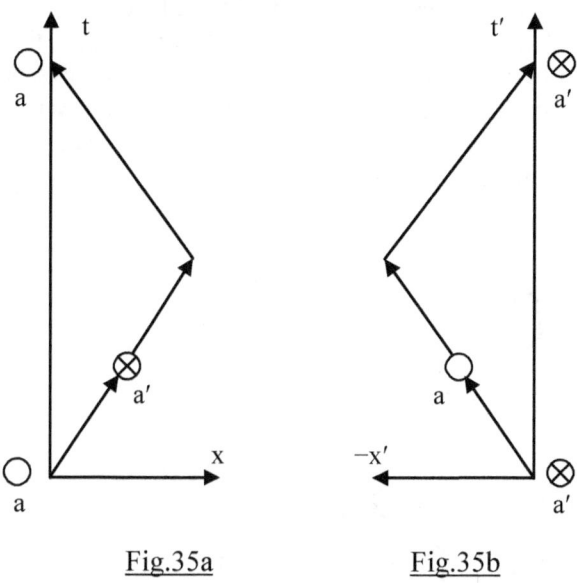

Fig.35a   Fig.35b

Now, it is the space twin that has a space-time path along his own time axis, while the earth twin moves relative to the space twin (with subsequent symmetric time dilation $t < t'$).

It will then be argued that the second space-time diagram is false, because the space twin "is non-inertial because he experiences accelerations". In other words, we have come full circle, and are back at the original assumption that "acceleration causes an asymmetry". So the space-time diagram hasn't actually proved anything, except represent the assumption of asymmetry in a graphical form. And as already proved, accelerations can only cause asymmetry while they act, so they cannot compensate for the symmetries during inertial phases. Thus, the problem remains unresolved.

Also, if the initial space-time diagram is accepted as true, then it implies the space twin is asymmetrically time dilated during inertial motion. This is because the space twin's space-time path can be subdivided into smaller sections, so that during his inertial motion, his clock is compared to the earth twin's clock. Of course, scientists and authors will try to argue that the clocks "cannot be directly compared" because a light signal "needs to travel from the twin's clock back to the earth". But I have proved this is a complete red-herring; systems of synchronised clocks at rest relative to the earth can be set up along the travelling twin's path. In this way, a direct comparison <u>can</u> be made. Machian Relativity provides an unambiguous resolution (Fig.36):

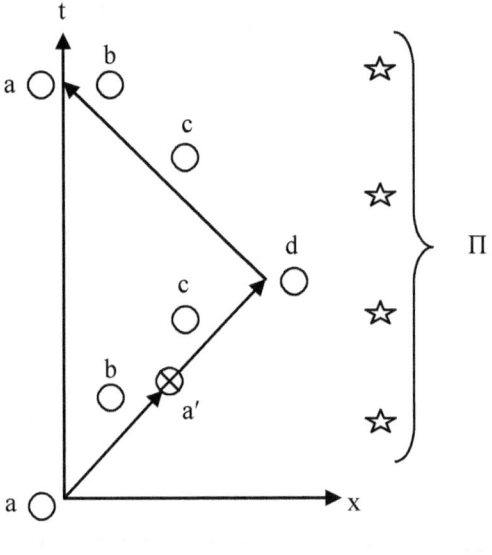

Fig.36

The space-time diagram can only be drawn the first way, with the earth twin at rest and the space twin moving. The external universe (the "Fixed Stars" Π) influences both twins, but the space twin more so. This is because the space twin has a higher absolute velocity through the Plenum of Mach Fields than the earth twin. Thus the space twin experiences physical and objective time dilation throughout the whole journey. He deduces this from comparing the reading on his clock (a′) with each stationary clock (a, b, c, d) as he passes by it. These compared clock readings can then be sent back to the earth twin in the form of digitised signals.

This is the simplest possible explanation for the asymmetry. And according to a principle of science (Ockham's Razor), the simplest explanation is usually the correct one.

Einstein's Initial Analysis

In Einstein's original paper (On The Electrodynamics of Moving Bodies, 1905) the results are often presented in an asymmetric way, by giving the equations for one system only. This is similar to the equations given by Lorentz for his Aether Theory, but this theory was asymmetric due to the Aether, whereas Einstein's wasn't. Einstein mentions that for an observer in uniform motion with a moving rod, a corresponding rod in the stationary system is similarly contracted, but does not give the equation. He then describes the situation for moving clocks, but only considers non-uniform motion (stopping and starting, circular or polygonal motion). The dilation is assumed to be asymmetric, without any explanation for it. This is done by giving the equations for one system only (as in Lorentz's theory). However, the expected dilation symmetry during continuous uniform motion is not considered.

Corollary: Rhetoric

Scientists and authors also use *rhetoric* to back up their arguments. For example, inertial clocks go slower when moving relative to each other because *"everything is relative"*, and *"Newton was wrong, because there is no absolute space or time"*. I say there is an absolute time, registered by clocks at rest relative to the average momentum of all matter (Mach's Principle). Therefore, clocks moving relative to this have absolute motion and measure relative time. So, in Machian Relativity, there is a mixture of relative and absolute time. So we need to say *"everything is relative; to the Plenum"*. It is said that *"Einstein's universe is democratic because different observers agree on the same things"*. A Machian universe is democratic too – all observers in inertial systems agree that the speed of light is constant and that their clocks and rulers change due to their motion relative to the Plenum. It is said that in Special Relativity *"the laws of physics are the same for different inertial observers*". Well, they are the same in Machian Relativity too (but in a more sensible way). It is said that "*Special relativity applies to inertial motion, but breaks down during acceleration, so General Relativity is needed*". But Machian Relativity applies to both inertial and accelerated motion, so it does not need to be saved by General Relativity. It is said that in Special Relativity "*you cannot rely on common sense*". But in Machian Relativity there is common sense because its equations obey the rules of logic and algebra. Also, conventional science tries to evade the issue by giving the "*so-called paradox*" a different name (e.g. "*the twin clock effect*").

Another example is "*no experiment conducted within an inertial system can detect its absolute motion*". But if clocks and rulers are changed by the system's absolute velocity, then they are equally affected, so there is no relative change between them. Therefore, such clocks and rulers need to be compared to other clocks and rulers external to the system (e.g. with a different velocity).

Science books also say that because of the finite speed of light, we cannot see a distant star as it is now, but as it was at some time in the past. In other words, there is no universal "now", but only a relative one, because "now" only has meaning in our immediate vicinity. While this is true, it is false to conclude from this that time dilation must be relative also. Also, separated clocks can be synchronised with each other, knowing their distances and the speed of light. So in this sense at least, there is a "universal now". Because of this, it is also claimed that simultaneity is relative, but this is not the relativistic simultaneity of Einstein's equations.

The Roundtrip Red-Herring

There is no actual requirement for the space twin to return to the earth for a comparison to be made. Instead, he can continue in his inertial motion from the earth indefinitely, according to Newton's First Law. The twins can observe each other as this happens, and compare their observations for asymmetries (previous chapter), or the space twin's clock can pass synchronised clocks at rest relative to the earth.

Conclusion

The simplest possible explanation for the asymmetric result in the Twins Paradox is that the travelling twin undergoes asymmetric time dilation for <u>every part</u> of his journey, including the inertial and accelerated phases. Thus the travelling twin observes the earth twin's time as a combination of time delays and *inverse dilation* (like in the Hafele-Keating Experiment when a clock is flown counter to the earth's rotation). If something looks like an elephant and walks like an elephant, then it is an elephant. And, if on reunion, the space twin looks like he has been ageing less during the whole journey, then that is what has been happening.

The problem with conventional science is that it retains the symmetry of Special Relativity, which then somehow has to generate an asymmetric outcome. There is no consensus among scientists as to which "explanation" is the definitive version; they cannot all be correct, as they would generate multiple dilation effects. What scientists are doing is bending the facts to save the theory, when it would be <u>better to change the theory</u>. A further objection to these methods is that none of them are used to explain the asymmetries experienced by moving clocks in Hafele-Keating experiments, where the relative rates of moving clocks are instead calculated by using the time dilation equations of Special Relativity in an asymmetric way. In this case, there can be no logical objection to applying the same method to the Twins Paradox, so that the twin ages asymmetrically for the whole journey.

A Brief History of The Twins Paradox

The Twins Paradox issue was first raised within the scientific community itself, by Paul Langevin in 1911. Scientists who supported Special Relativity, including Einstein, proposed a variety of "solutions" to circumvent the problem, and thereby save the theory. But, there has never been any consensus on a definitive version. Other scientists were more critical of Special Relativity and instead preferred alternatives, such as Lorentz's Aether Theory (which explained the travelling twin's asymmetric ageing in terms of physical time dilation for every part of the journey, due to Maxwell's Luminiferous Aether defining a state of absolute rest).

Over the century since the theory was published, opinions within the Scientific Establishment contrary to the Einsteinian Paradigm have been silenced. There was a brief revival of the controversy in the 1970s with the publication of Science at the Crossroads by Herbert Dingle. Debate on the issue is now conducted outside the scientific community by people who are branded as "cranks" with "crackpot theories" by scientists and authors in the media. These ad hominem attacks have been so effective that even non-scientists are copying them in Internet forums. Scientists occasionally lampoon the more obviously erroneous theories, while maintaining a dignified silence on the more convincing alternatives. However, the majority of people who study physics in universities do not end up working in them, so there could be many scientifically trained non-academics working on the issue. Their denigration perhaps also explains how criticism of Einstein's theory has been expunged from within the Scientific Establishment – no professional scientist would dare criticise it and risk being branded a crank.

There is also a trend where a spurious history of Relativity is being written. The contributions of other scientists to the subject before Einstein published his theory are glossed over. Einstein is portrayed to the public as a lone genius who arrived at his theories by "pure thought".

Even the Twins Paradox is being surreptitiously transformed into something else called the "Twin Paradox". This is "the paradoxical result that a travelling twin will be younger than his brother when he returns to earth". Now, this asymmetric outcome might be strange and unusual, but it is certainly not a paradox. The paradox is in Special Relativity's inability to properly explain this asymmetry with its contradictory prediction that the twins would be younger than each other when reunited (because it is based on purely relative motion). Scientists have tried to fudge the asymmetry with a variety of incompatible "explanations" to "rectify" the theory, but none have been experimentally proven.

KAPPA

VARIOUS REFUTATIONS OF THE EQUIVALENCE PRINCIPLE

PROPOSITION I: The Equivalence Principle violates Newton's Second Law of Motion when applied to rotating systems.

Consider an observer in a rotating system. In this system, objects apparently experience an outward force. For example, if a stone is attached to a length of string, the string is under tension if the other end is held in place. According to the Equivalence Principle, the observer can consider the system as being at rest in the centre of an outward (centrifugal) gravitational field, with the external universe rotating about the system.

However, there is a problem. Newton reasoned that orbital motions (such as the moon or a satellite orbiting the earth) arise from attractive forces. Thus, if the rotating observer uses Einstein's Equivalence Principle again, he must conclude that the circular motion of the external universe is due to an attractive (centripetal) gravitational force. But this contradicts the previously postulated outward (centrifugal) gravitational field. So, there is a paradox, as the Equivalence Principle predicts contradictory forces. This also means Newton's Second Law of motion has been violated, if (say) the forces predicted for the external universe are considered to act on the stone. Therefore, the Equivalence Principle is refuted by this argument.

In addition, let the mass of the observer be insignificant compared to the mass of the system. If the observer is released from the system, he observes that he moves inertially from the system at a tangent. Again, such motion is not consistent with the postulated field, because it does not accelerate the observer.

Therefore, like observers external to the system, a rotating observer concludes that centrifugal effects actually arise due to the centripetal force (that maintains circular motion) and *centrifugal inertia* – the tendency of a released body to recede inertially from a rotating system. I say this inertia is relative to the Plenum (Fixed Stars). On this basis, rotating observers must conclude that the rotation of the external universe is apparent (relative) due to there being no cause inherent to the supposed rotation, and that instead the cause is within the system itself, which I say is its *absolute rotation relative to the external universe.*

A true <u>centrifugal force</u> arises between mutually repelling bodies, such as by electric repulsion. Let a charged body be connected by (say) string to a rotating system with the same charge polarity. The tension in the string must now provide a centripetal force that overcomes both the body's centrifugal inertia and the centrifugal force. When released, the body moves (relative to external observers) from the system with a motion that combines its initial tangential inertia with an imparted radial acceleration from the repulsive force. Observers on the rotating system also observe this, along with the additional relative rotation of the external universe.

A rotating system cannot consider itself as being at rest for other reasons. Consider the earth – atomic clocks flown in the direction of the earth's daily rotation work slower than the same clocks flown against the earth's rotation. This difference arises from the earth's absolute rotation relative to the external universe. Other authors and scientists might claim (like Mach) that this would also happen if the external universe were caused to rotate about the earth instead. In response to this, I say that:

a) If the external universe (the "Fixed Stars") were caused to rotate, this would require a mutual interaction with other matter, which would possess an equal and opposite angular momentum. Thus, there would be no total angular momentum around the earth. Therefore, the earth would experience no centrifugal effects, nor any asymmetry in the behaviour of the atomic clocks.

b) No credible theory of mechanics can explain the forces

required to maintain the rotational motion of external matter about a stationary earth. For example, a postulated induced gravitational field would be required to increase with distance, but actual gravitational fields decrease with distance.

PROPOSITION II: The Equivalence Principle generates a circular argument when extended to acceleration by applied forces in free space.

In Einstein's theory of General Relativity, the Equivalence Principle is a relativity of acceleration. A system falling freely in the earth's gravitational field experiences no acceleration effects in its immediate vicinity, and is considered as being inertial. Likewise, a system undergoing acceleration by applied forces in free space, experiences non-inertial effects. Such a system is said to be equivalent to a system at rest on earth, and is considered to be at rest in an induced gravitational field.

The first purpose of the Equivalence Principle is to explain the observational fact that all objects in the gravitational field of a body (such as earth) fall at the same rate, regardless of their different masses. If a system in free space is accelerated by applied forces, it feels acceleration effects, and relative to it all external bodies move with the same acceleration, independently of their different masses (in Newtonian mechanics these are called "fictitious forces"). Locally, a system at rest in a gravitational field experiences the same effects. Einstein says this is because it is "equivalent" to the accelerating system in free space. This is said to "explain" the equivalence of gravitational mass with inertial mass; $m(g) = m(i)$. Because of this, in a localized region of a spherical gravitational field, objects move inertially relative to each other. So in a falling system objects obey Newton's First Law, causing the effect of zero-g. According to Einstein, such a falling system is equivalent to an inertial system in free space.

So, Einstein explains the equivalence of gravitational and inertial masses by comparing systems in gravitational fields to those in free space. Now, the second purpose: Einstein extends his argument by considering the reverse situation - comparing what happens in free space to what happens in a body's gravitational field. This is Einstein's attempt to abolish Newton's Absolute Space and instead explain the asymmetry between inertial and accelerated systems in terms of his Equivalence Principle. That is, a system accelerating in free space can consider itself at rest in a gravitational field in which external objects "fall". The gravitational field is somehow induced by applied forces and is not associated with any source, such as a planet's mass. So, all inertial external objects "fall" at the same rate in the induced field.

Now for the problem: how do we know bodies "fall" at the same rate in this postulated field? Answer: because objects fall at the same rate in the gravitational field of an actual body, such as a planet. And how do we know this happens? Because they are equivalent to inertial objects in free space - which are themselves "falling" in an induced field! What Einstein has done is generate a circular argument in which two things are explained in terms of each other.

Einstein criticised Newton for invoking his Absolute Space that "acts on matter but which cannot be acted on itself". But Einstein has:
a) Filled the "free space" around accelerated systems with induced gravitational fields, making it no longer free.
b) Invoked the existence of induced gravitational fields, which (unlike the gravitational fields of celestial objects) have no mass acting as a source and are therefore arbitrary. That is, objects falling in such an induced field have no source of mass with which to mutually interact with, and therefore violate Newton's Third Law of action and reaction. In addition, the postulated field must be infinite in extent so that the whole external universe "falls" in it. But it is not logical to believe that an accelerated system could have this effect on the whole universe, especially when its observers know that their system is mutually interacting with other matter, such as rocket gases.

It must be concluded that if systems falling in gravitational fields are inertial (according to the Equivalence Principle), then to avoid

Glyn Phillips

the circular argument the reverse situation cannot apply. That is, accelerated (non-inertial) systems are not different from inertial ones due to "being at rest in induced gravitational fields", but because of some other explanation - *which I say is Mach's Principle*. That is, the acceleration effects experienced by non-inertial frames are generated by the average momentum of all matter, which acts across all space due to the Plenum (the combined effect of the Mach Fields of all matter in the Universe). Therefore, in the theory of Machian Relativity, "space" does not act on matter, but rather matter acts on other matter, thereby answering this criticism of Newton by Einstein. In addition, Einstein quoted the experiments of Eotvos, as a "proof" of his Equivalence Principle. These showed no difference in gravitational and inertial mass. However, the evidence for this already existed in classical physics, because the orbital speeds of the planets depend on the mass of the sun and their distance from it, but not on the mass of the planet itself. Newton formulated his equations to allow for this. What Eotvos did was to draw the same conclusions from laboratory experiments as Newton did from astronomical observations over two centuries earlier.

PROPOSITION III: The Equivalence Principle can be refuted by introducing additional non-inertial systems.

Consider an inertial system in free space, which is then accelerated by applied forces, such as by firing rockets. According to the Equivalence Principle of General Relativity, such a system can consider itself at rest in a gravitational field in which other objects fall. Such objects would be everything else in the universe.

Now consider a second inertial system, identical to the first. When the first system accelerates, the second system accelerates in the other direction. According to General Relativity, this second system can also consider itself to be at rest in a gravitational field, this time acting in the opposite direction to the field of the first system. It can also consider external objects to fall in this field, but in the opposite direction in comparison to the first system. In other words, both systems can be considered to have generated equal and opposite gravitational fields. Now comes the problem - both fields

119

must cancel each other to produce a zero resultant gravitational field. Thus, the field of the first system is cancelled by the field of the second system, and vice versa. Thus, neither system can now be considered to be at rest, as there is now no resultant gravitational field. On the basis of this argument, *Einstein's hypothesis of accelerating systems being at rest in induced gravitational fields is refuted*, and instead another explanation must be found to explain the non-inertial effects experienced by accelerating systems. Such an answer is provided by Machian Relativity, where inertial systems move inertially relative to the Plenum unless acted on by a resultant force.

Corollary

A similar problem arises when the effects of a system's mutual interactions during acceleration are considered. For example, if it is accelerated by rockets, the system mutually interacts with its rocket gases according to Newton's Third Law. So, when the system accelerates in one direction, its rocket gases accelerate in the other. Now, if the Equivalence Principle applies to the system, it must also apply to the rocket gases. This means that the rocket gases can be considered at rest in a gravitational field, acting in the opposite direction to the system's field. And because the mass of the expelled gas is less than the rocket's, the acceleration of the gas is correspondingly higher. By application of the Equivalence Principle, the rocket gases experience a similarly higher G-field. As these induced G-Fields act uniformly throughout space affecting external matter, it must also affect the system being accelerated. By deduction the system must experience the vector sum of its own induced field and the field of its expelled gases. As its own field has a much lower magnitude, this means the system must experience a net G-Field in the same direction as that experienced by the gases. This contradicts the direction of the induced field initially postulated by the Equivalence Principle, thereby refuting it. Another problem is that such induced fields act instantly across all space, violating the standard idea that the speed of light is a limiting factor.

PROPOSITION IV: The Equivalence Principle contradicts the standard Twins Paradox explanation.

Consider two twins on the earth. One goes on a journey from the earth, while the other stays at home. The traveller is accelerated, then coasts inertially. The traveller is then decelerated, reversing his motion. He then coasts inertially back to earth and is brought to rest.

On reunion, the travelling twin has aged less than the earth twin. According to the standard explanation, this asymmetry arises because the travelling twin has experienced forces, and is therefore "non-inertial", while the earth twin is inertial because he has not experienced these forces. As explained earlier, this explanation is false because the accelerations cannot compensate for the periods of inertial motion. However, there is a far more serious objection to the standard explanation, which is that the earth twin is not actually inertial if the Equivalence Principle is invoked. The earth twin is at rest in the earth's gravitational field because he is on the earth's surface. According to Einstein's Equivalence Principle, systems at rest in the gravitational fields of massive bodies are equivalent to systems being accelerated by applied forces in free space. So, if accelerated systems are non-inertial, then the earth twin must also be non-inertial. This paradigm of standard physical science therefore leads to the conclusion that both twins must be non-inertial. So, the conventional argument that the travelling twin ages less than the earth twin because "he feels forces and is non-inertial, while the earth twin does not" is invalidated and replaced by an ambiguity.

And the earth twin always feels forces, while the space twin only feels forces during the parts of the journey when he accelerates. So, on this basis, conventional science leads us to conclude that the earth twin should instead age less than the space twin.

My theory of Machian Relativity properly resolves this – the travelling twin ages less than the earth twin for all parts of the journey (inertial and accelerated), due to his absolute motion relative to the external universe.

PROPOSITION V: The Equivalence Principle reverses the expected asymmetries in Relativity

Einstein concluded from Special Relativity that due to the earth's rotation, a clock at the equator would have a slower rate than one at either the north or south poles. This has been verified with atomic clocks on planes and satellites, although Machian Relativity provides a more rigorous explanation than Special Relativity.

Now apply the Equivalence Principle. A clock on the earth feels forces, while one on the orbiting satellite does not because it is in free-fall. So the earth clock is non-inertial, and the satellite clock is inertial.

Now consider the Twins Paradox. Supposedly, the travelling twin physically ages less than the earth twin, because the traveller "experiences forces and so is non-inertial".

Now combine the Equivalence Principle with the Twins Paradox: the earth clock is non-inertial and thus should have a slower rate than the inertial satellite clock. But, this contradicts what actually happens.

The proper explanation is provided by Machian Relativity. The satellite clock has a greater absolute motion relative to the "Fixed Stars" than the earth clock, and so has a slower rate. And Machian Relativity derives this asymmetric result from first principles, whereas in Special Relativity Einstein merely assumed it from symmetric equations. The same applies to the Twins Paradox – the traveller ages less than the earth twin because he has a greater absolute motion relative to the Plenum ("Fixed Stars") than the earth twin. And the slower ageing occurs during the motion, just like the satellite clock has a slower rate during its orbital motion.

Corollary

Experiments on atomic clocks on planes prove that they physically work slower than clocks on the earth, and that this happens during the motion itself. This is due to Machian Relativity. This physical change will also apply to orbiting clocks on satellites. The Equivalence Principle claims that such gravitational free-fall is an inertial process; therefore Special Relativity should apply to these clocks. But Special Relativity was originally formulated for inertial clocks moving in straight lines, and it predicts reciprocal time dilation between them. However, the physical changes in the orbiting clocks (relative to the earth) prove that their time dilations cannot be reciprocal when they are moving relative to each other – if they are asymmetric relative to earth clocks, then they must be asymmetric relative to each other. For example, let two systems of clocks have different orbits, so that they are in relative motion. Once other factors have been factored out (e.g. time delays, gravitational dilation, earth's rotation), then both systems cannot have reciprocal time dilation relative to each other – it must be asymmetric.

Also, let two clock systems share the same orbit (to eliminate their gravitational differences). They orbit in opposite directions, and are compared each time they pass each other (to eliminate time delays). As a result, they have essentially the same time dilation as each other, and so measure no difference in the other's clock readings. Thus, they do not measure each other's clock readings to change symmetrically according to their relative motion as required by Special Relativity.

PROPOSTION VI: The Equivalence Principle cannot apply to Systems in Free Space

It is known from experiments that a clock's rate is determined by its gravitational potential energy. Atomic clocks run at a faster rate when at a higher altitude. This gravitational time dilation was predicted by Einstein using his General Relativity, and occurs along with the more usual dilation by high speed.

Now consider a system in deep space far from massive bodies. If it is accelerated by applied forces, then by the Equivalence Principle observers in the system can consider themselves at rest in an induced gravitational field in which the external universe "falls". So, General Relativity (supposedly) applies to this system too. If the system has clocks on the axis along which it is accelerated, then clocks at the front are at a higher gravitational potential than those at the back. So clocks at the front have a faster rate, while clocks at the back have a slower rate.

However, although this tells us how the rates of the accelerated clocks change in relation to each other, it gives no indication of how these clocks change relative to unaccelerated ones. In the proper gravitational field of a massive body this is not a problem – clocks at infinite distance are unaffected because there is no gravitational field, and the gravitational dilation increases the closer they are to the body.

For example, if the rearward clock is defined as the origin or reference, then during the acceleration it undergoes a normal asymmetric time dilation determined by its instantaneous absolute motion relative to the Plenum. Thus relative to observers at rest, the forward accelerated clocks must change by a combination of motional and gravitational dilation. However, if the forward clock is defined as the origin, then it will have only motional dilation relative to observers at rest. So, the physics of the system relative to stationary observers is affected by an arbitrary choice of reference clock. Therefore, the Equivalence Principle gives an ambiguous result for acceleration in free space and must be rejected.

Corollary I

A further problem lies with relativistic simultaneity. This requires the forward clock of a moving system to have a lower reading than the rearward clock in relation to static observers. However, during acceleration, the Equivalence Principle predicts that the forward clock runs at a faster rate than the rearward one. So when the acceleration stops, the forward clock of the moving system has a higher reading instead. Machian Relativity avoids these problems because clocks only have equal motional dilation during acceleration, while gravitational dilation and relativistic simultaneity do not occur.

Corollary II

Like the clocks in an accelerating system, clocks in the external universe are similarly influenced by the induced gravity field. So, external clocks approaching the system are at a higher gravitational potential and thus a faster rate (lower gravitational dilation), while receding clocks have a slower rate (higher gravitational dilation). However, according to the Einsteinian Paradigm, because observers in the accelerating system are "at rest", external clocks must also be affected by motional time dilation. Therefore, when the acceleration stops, the induced gravity field must vanish. So, gravitational dilation will stop, and the external universe will only have motional dilation relative to the (now) inertial system (e.g. the reciprocal dilation of Special Relativity). However, as pointed out by Herbert Dingle in his book, because of the different gravitational time dilations previously experienced by the external universe, the initial synchronisations of all clocks throughout the universe will have been altered by the induced gravity field. Time itself would have passed at different rates in the universe. So the Equivalence Principle leads to highly implausible consequences. No, Machian Relativity is far better. There are no induced gravity fields or reciprocal time dilation. The accelerating system experiences non-inertial effects and time dilation because of the external universe. The external universe changes the system but not vice-versa, because the mass of the universe is greater than the system. Einstein's demands for equivalence are futile. The Equivalence Principle can state that observers in non-inertial and inertial systems are "at rest", but such systems are not equivalent in other terms; such as their different inertial effects or their asymmetric time dilations.

Einstein says a gravitational field is "produced" in an accelerating system, but this is false. Although magnetism is produced by electric currents, the magnetic fields objectively exist for observers at rest or moving with the charges. And parallel electric currents mutually interact with each other. But the external universe that "falls" in the induced field is not mutually interacting with other matter, thus the field cannot exist for such matter. However, the accelerating system is interacting with other matter (rocket thrust), so it is an objective fact the system is moving, but the external universe is not.

Scholium

The ancients believed that all celestial bodies, including the stars, rotated around a stationary earth. This geocentric model was overthrown by the heliocentric model of Copernicus and Galileo – the earth and other planets orbited the sun, which provided a better explanation of how the planets were observed to move relative to the earth. These motions were eventually explained by Newton's Laws of Motion and Universal Gravity. So the observed rotation of the stars is due to the earth's inherent daily rotation. However, the Equivalence Principle (if true) undermines this. Thus, the choice is not between a geocentric viewpoint and a heliocentric one, but both are valid, because observers on the rotating (and orbiting) earth can consider themselves "at rest". The geocentric system of the ancient world relied on concentric celestial spheres and wheels within wheels. However, the "relativistic geocentric system" of the Equivalence Principle is achieved through Einstein's induced gravity fields. So, it is impossible to objectively know whether the earth is moving or stationary, because everything is relative and there is no state of absolute motion or rest. Thus absolute rotation can have no objective meaning.

Now, the moon orbits the earth monthly due to its gravitational attraction, and its observed daily motion is apparent, due to the earth's inherent rotation. However, according to the Equivalence Principle, the earth's daily rotation is not inherent. Instead, rotating frames are "at rest" in an induced centrifugal gravitational field. This supposedly explains effects such as the earth's equatorial bulge. But the outward field contradicts the attractive force producing the moon's monthly orbit. In addition, the moon's apparent daily orbit would require an additional gravitational attraction which also contradicts the postulated centrifugal field.

Neither can this explain the daily rotation of the stars. There is no plausible force to maintain such supposed around the earth, because we know the earth's gravity decreases with distance. And the field postulated for the daily rotation of celestial bodies contradicts the field that would be required to explain the retrograde motion orbits of planets around a supposedly stationary earth. All these supposed fields are entirely arbitrary and inconsistent with each other (one could argue retrograde motion arises due to the planets having additional motion around the sun, but if the planets do this, then the earth should too). Thus the Equivalence Principle reintroduces the geocentric universe, although in a relative sense only. But like the Ptolemaic Paradigm, Einstein's Paradigm is bedeviled with problems by ascribing physical causes to purely apparent effects (retrograde planetary motion and daily celestial rotation).

My various arguments refute these ambiguities[7]. We know that when two bodies interact, the greater mass moves less than the lesser mass. Thus the earth moves because it is less massive than the sun, and this motion is absolute. Such motion also manifests itself in stellar parallax and stellar aberration. Thus we must replace the Equivalence Principle with the Principle of Machian Inertia:.

> Bodies remain at rest or move in a straight line at constant speed, relative to the average momentum of all matter, unless acted on by a resultant force (resulting from mutual interactions between bodies).

---

[7] Not only does this book show how Mach's Principle explains space-time asymmetries during inertial motion, but that Mach's original interpretation of his idea (that acceleration is relative not absolute) is erroneous. Rotating the stars around an observer *will not cause* centrifugal effects, as this would require interaction with other matter having an equal and opposite angular momentum, thereby cancelling any effect on the observer. Thus, effects are relative to the *average momentum of all matter*, not just the "Fixed Stars" per-se, and this defines a true state of absolute rest.

LAMBDA

THE EFFECTS OF TIME DELAYS ON MEASUREMENTS

So far, only *physical* changes between systems have been considered. This has been achieved by a direct comparison of clocks and rulers in each system as they pass in the immediate vicinity of each other. Thus apparent effects, such as time delays are factored out. So, if an observer sees a clock in the other system directly opposite him, he sees the clock how it actually is at that instant. This chapter will consider how a clock's rate appears to an observer when they continuously move relative to each other. Because the distance between the clock and observer continuously changes, the time delay for light to reach the observer from the clock continuously changes. This can take the form of an image of the clock (if illuminated by a light source in any relative motion), or clock readings emitted from the clock as a digitised signal. This chapter is necessary to refute the erroneous idea that Time Delays can produce physical effects (such as in the Twins Paradox).

Calculation: Effects on Observed Dilation for a Clock receding from an Observer at Absolute Rest

Consider the stationary and moving systems S_α and S_B as before. Clocks in the stationary system are synchronised so that they instantaneously share the same reading. Clocks in the moving system are synchronised according to the equations of Relativistic Simultaneity. So, when the origins of both systems coincide, the clock at $x_\alpha 1 = 0$ reads $t_\alpha 1 = 0$, and the clock at $x_\alpha 2$ reads $t_\alpha 2 = 0$. In S_B, the clock at $x_B 1 = 0$ reads $t_B 1 = 0$, and according to Relativistic Simultaneity, the clock at $x_B 2$ reads:

$$t_B 2 = -v_\alpha x_\alpha 2 / c^2 \sqrt{(1 - v_\alpha^2/c^2)} = -v_\alpha x_\alpha 2 / c^2 \gamma = \Sigma_B \{x_\alpha\}$$

This Relativistic Simultaneity is also an objective fact for the moving system S_B. Because of Lorentz contraction in S_B, the corresponding measurement made by observers in S_B is $x_B2 = x_\alpha2/\gamma$ so that:

$$t_B2 = -v_\alpha x_B2 / c^2 = \Sigma_B\{x_B\}$$

Light spheres from sources in both systems have absolute motion relative to S_α. That is, their wavefronts Ψ_α and Ψ_B expand uniformly relative to S_α. This has been previously demonstrated due to the interaction between the Mach Fields of S_A and S_B. Therefore, in the frame S_α the time delay for clocks $x_\alpha2$ and x_B2 is, for an observer situated at the origin O_α, given by:

$$D_\alpha = x_\alpha2 / c$$

For the observer in S_α, he sees clocks in his own system to have the same rate as his own, but with a time delay due to their distance. The same observer also sees clocks in S_B at the same distance with the same time delay. So at $t_\alpha = D_\alpha$ the observer sees the clock at $x_\alpha2$ read $t_\alpha = 0$, and at the same instant he sees the clock x_B2 read $t_B2 = \Sigma_B$ (according to relativistic synchronisation).

However, for clocks in S_B, there are the added effects of their time dilation and that their distance relative to this observer is continuously changing, due to their relative motion. When the clock at x_B1 reaches $x_\alpha2$, its reading relative to clocks $x_\alpha1$ and $x_\alpha2$ is given by the time dilation equation:

$$t_B1 = \gamma\, t_\alpha2$$

However, because of the time delay, the stationary observer doesn't see this until his clock reads:

$$t_\alpha1 = t_\alpha2 + x_\alpha2 / c$$

Because $x_\alpha2 = v_\alpha.t_\alpha2$, we have for this reading:

$$t_\alpha1 = t_\alpha2 + (v_\alpha.t_\alpha2 / c) = t_\alpha2 (1 + v_\alpha / c)$$

If the moving clock (x_B1) travels a distance $x_\alpha = v_\alpha.t_\alpha3$ in the system S_α, its reading due to time dilation is:

$$t_B1 = \gamma.t_\alpha3$$

So due to time delays, the stationary observer sees the moving clock with this dilated reading when his clock reads:

$$t_\alpha1 = t_\alpha3 + (v_\alpha.t_\alpha3 / c) = t_\alpha3 (1 + v_\alpha/c)$$

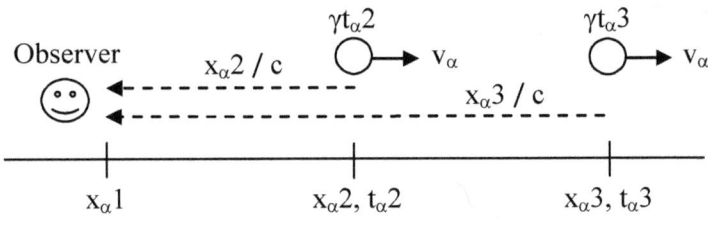

Fig.37

Now, the <u>true rate</u> of the moving clock (without time delays) is:

$$\gamma \text{ (true)} = \gamma (t_\alpha3 - t_\alpha2) / (t_\alpha3 - t_\alpha2) = \gamma$$

But because of time delays, the *observed rate* relative to the observer is:

$$\gamma \text{ (obs)} = \gamma (t_\alpha3 - t_\alpha2) / \{(t_\alpha3 - t_\alpha2) (1 + v_\alpha/c)\}$$
$$= \gamma / (1 + v_\alpha/c)$$
$$= \sqrt{(1 - v_\alpha^2/c^2)} / (1 + v_\alpha/c)$$

Hence, the observer sees the receding clock with a rate determined by its physical time dilation <u>and</u> an extra reduction factor $1/ (1 + v_\alpha/c)$, which is *apparent in nature* due to its increasing time delays, given by:

$$\gamma \text{ (app)} = 1/ (1 + v_\alpha/c)$$

Glyn Phillips

Corollary: Effects on Observed Dilation for a Clock approaching an Observer at Absolute Rest

What if instead, the moving clock is approaching the stationary observer?

Let the clock be at distance $x_\alpha 2$. The time delay to the stationary observer is $x_\alpha 2/c$. Let the clock reading be $t_B 2$. It then travels a distance $-v_\alpha (t_\alpha 3 - t_\alpha 2)$. So the time delay to the observer is now:

$$D_\alpha = x_\alpha 2/c - v_\alpha (t_\alpha 3 - t_\alpha 2)/c$$

So at $x_\alpha 2$, the clocks read $t_\alpha 2$ and $t_B 2$ respectively, and at $x_\alpha 3$ they read $t_\alpha 3$ and $t_B 3$ respectively. So the true rate is:

$$\gamma \text{ (true)} = (t_B 3 - t_B 2) / (t_\alpha 3 - t_\alpha 2)$$

The observer gets the readings ($t_B 2$ and $t_B 3$) of the moving clock at $(t_\alpha 2 + x_\alpha 2/c) = t(\text{obs}1)$ and $(t_\alpha 3 + x_\alpha 2/c - v_\alpha (t_\alpha 3 - t_\alpha 2)/c) = t(\text{obs}2)$. So the observed rate is:

$$
\begin{aligned}
\gamma \text{ (obs)} &= (t_B 3 - t_B 2) / \{t(\text{obs}2) - t(\text{obs}1)\} \\
&= (t_B 3 - t_B 2) / \{t_\alpha 3 - t_\alpha 2 - v_\alpha (t_\alpha 3 - t_\alpha 2)/c\} \\
&= (t_B 3 - t_B 2) / \{(t_\alpha 3 - t_\alpha 2)(1 - v_\alpha/c)\} \\
&= \gamma \text{ (true)} / (1 - v_\alpha/c)
\end{aligned}
$$

Hence, the observer sees the approaching clock with a rate determined by its physical time dilation and an extra apparent factor due to its decreasing time delays:

$$\gamma \text{ (app)} = 1 / (1 - v_\alpha/c)$$

The use of a moving clock receding from the origin of the stationary system allows relativistic simultaneity effects to be factored out. For other clocks, the same apparent dilation occurs, but the clock readings are correspondingly shifted according to relativistic simultaneity.

131

Corollary: How an Absolutely Stationary Clock's Dilation appears to a Receding Observer with Absolute Motion

Next, we must consider how a clock in S_α looks relative to the observer in S_B. This is because relative to the observer, clocks in S_α have relative motion and so their time delays continuously change. When the origins coincide, the clock in S_α and its corresponding clock in S_B both read zero: $t_\alpha 1 = t_B 1 = 0$. When $t_\alpha = t_\alpha 2$, the clock in S_B has travelled a distance $v_\alpha t_\alpha 2$. But where is the moving observer when he sees the stationary clock at the origin of S_α read $t_\alpha = t_\alpha 2$, and what does his own clock read? When this clock reads $t_\alpha = t_\alpha 2$, its wavefront then travels to the moving observer. Relative to the x–axis of S_α, the position of the moving observer is given by $x_\alpha 2 = v_\alpha t_\alpha 2$. The position of the wavefront is:

$$x_\alpha = c\,(t_\alpha - t_\alpha 2) \quad \{\text{the wavefront does not exist prior to } t_\alpha 2\}$$

This wavefront reaches the moving observer when:

$$v_\alpha t_\alpha 3 = c(t_\alpha 3 - t_\alpha 2)$$
$$t_\alpha 3\,(v_\alpha - c) = -\,ct_\alpha 2$$
$$t_\alpha 3 = ct_\alpha 2\,/\,(c - v_\alpha)$$

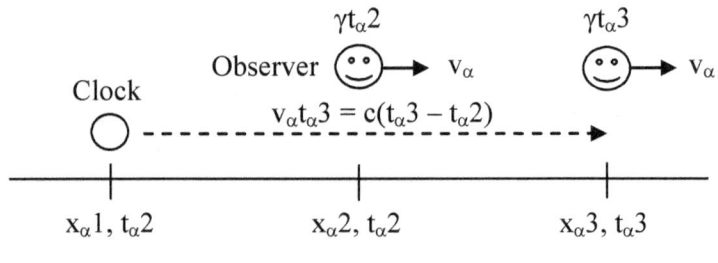

Fig.38

Therefore, the moving observer sees the stationary clock reading $t_\alpha 2$ when clocks in S_α read $t_\alpha 3$. Because of time dilation in S_B, the corresponding reading of the moving observer's clock is:

$$t_B3 = \gamma \cdot t_\alpha 3 = \gamma \cdot t_\alpha 2 / (1 - v_\alpha / c)$$

Thus for the moving observer in S_B, this gives:

$$t_\alpha 2 = t_B 3 (1 - v_\alpha/c) / \gamma$$

Or, the rate of the clock in S_α, as observed by the observer in S_B, is:

$$\gamma (obs) = (t_\alpha 2 - t_\alpha 1) / (t_B 3 - t_B 1) = (1 - v_\alpha / c) / \gamma$$

Where: $1 / \gamma$ is the true *anti–dilation* (inverse dilation) of clocks in S_α relative to S_B. The other factor is the apparent rate due to increasing time delays, arising from the recession of the stationary clock as seen by the moving observer in S_B:

$$\gamma (app) = (1 - v_\alpha / c)$$

So the anti–dilation tends to give the stationary clock an observed faster rate, while the apparent factor tends to give the stationary clock an additional slower rate.

Corollary: How an Absolutely Stationary Clock's Dilation appears to an Approaching Observer with Absolute Motion

Similarly[8], the moving observer will see an approaching stationary clock with an observed rate given by:

$$\gamma (obs) = (1 + v_\alpha/c) / \gamma$$

Where $1 / \gamma$ is the anti–dilation of clocks in S_α relative to S_B, which is the true rate. The other factor is the apparent rate due to time delays:

$$\gamma (app) = (1 + v_\alpha/c)$$

---

[8] S_α is at absolute rest relative to the average momentum of S_A and S_B. By including all other matter in the universe, S_α is at absolute rest relative to the average momentum of the Universe.

Corollary: the Overall Cancellation of Apparent Dilations relative to an Observer at Absolute Rest

The various apparent dilations of a clock seen by an observer (due to their different inertial motions and time delays), have been determined. However, is it possible for time delays to have a residual effect on clock readings, when a moving clock (or twin) returns to the earth? And is it possible for such a residual effect to hide the paradoxical and unphysical nature of Special Relativity?

On the outward journey, the earth twin sees the moving twin to have time dilation modified by time delays. According to Special Relativity, the moving twin observes the earth twin to behave in exactly the same way. When the moving twin turns around, he sees an immediate change in the earth's relative motion. But, the earth twin does not see a correspondingly immediate change for the moving twin. Supposedly, this difference means the twins see different things on reunion. Firstly, the time delay effects seen by the earth twin, for the space twin's return journey, cancel the corresponding effects for the outward journey, so that the earth twin sees the space twin to have only time dilation on reunion. Secondly, the moving twin supposedly sees the earth twin with much faster ageing on the return journey than the corresponding effect seen by the earth twin. Thus, the time delay effects seen by the space twin for the outward and return journeys do not cancel, so that he sees the earth twin with overall advanced ageing on reunion. This supposedly hides the paradoxical symmetry of Special Relativity. However, we might also expect the overall effects of apparent dilations (as seen by either twin / observer) to vanish on reunion, because the overall time delay (separation) has similarly vanished. So what actually happens?

Consider the clock in S_B moving from the origin of S_α. Relative to S_α, it travels a distance $x_\alpha = v_\alpha t_\alpha$. At this point x_α in S_α, the clock then turns around. According to Machian Relativity, if a wavefront is emitted from the clock when this happens, it has absolute motion relative to S_α (which is at the average momentum), so it takes a further time x_α / c to reach the observer at the origin of S_α (this also occurs in the standard explanation using Special Relativity, but is instead assumed rather than proved). In this time, the clock has travelled a distance $v_\alpha . x_\alpha / c$ back to the observer. Thus, the actual position of the clock relative to the observer (when the observer sees the clock turn around) is:

$$x_\alpha 2 = x_\alpha 1 - x_\alpha 1 . v_\alpha / c = x_\alpha 1 (1 - v_\alpha / c)$$

The duration of recession, seen by the stationary observer, is:

$$t_\alpha 2 = t_\alpha 1 + x_\alpha 1 / c = t_\alpha 1 + v_\alpha . t_\alpha 1 / c = t_\alpha 1 (1 + v_\alpha / c)$$

The duration remaining for the approach, as seen by the stationary observer, is:

$$t_\alpha 3 = t_\alpha 1 - x_\alpha 1 / c = t_\alpha 1 (1 - v_\alpha / c)$$

[This follows because we require $t_\alpha 2 + t_\alpha 3 = 2t_\alpha 1$, which is the total duration of the journey relative to the stationary observer].

Now, we expect the observed dilations of the moving clock (the product of its apparent and physical dilations), as seen by the stationary observer, to combine in such a way, that on reunion only the physical dilation remains (γ), while the apparent dilations (Ξ) for the outward and inward journeys are nullified. Thus:

$$\gamma . \Xi(\text{in}) . t_\alpha 1 (1 - v_\alpha / c) + \gamma . \Xi(\text{out}) . t_\alpha 1 (1 + v_\alpha / c) = 2\gamma t_\alpha 1$$

So:

$$\Xi(\text{in}) (1 - v_\alpha / c) + \Xi(\text{out}) (1 + v_\alpha / c) = 2$$

The previously calculated apparent rates for a moving clock satisfy the above condition. QED: Apparent rates due to time delays vanish when the moving clock returns to the stationary observer. Thus, the only remaining effect seen by the observer on reunion is the true dilation of the moving clock due to its absolute motion.

Corollary: the Overall Cancellation of Apparent Dilations relative to an Observer with Absolute Motion

It can also be shown that the apparent dilations (Ξ) of a stationary clock relative to a moving observer are similarly nullified on reunion. During the moving twin's turnaround, he immediately sees the stationary clock change direction. Thus, for the moving twin, the outward and inward phases have equal durations. Thus $t_B(out) = t_B(in) = t_B 1 = \gamma t_\alpha 1$. The total duration for the moving observer is $2t_B 1$. The anti-dilation of the stationary clock is $1 / \gamma$. We expect the stationary clock to have only its anti-dilation on reunion, so that it reads $2t_B 1/\gamma$ ($= 2t_\alpha 1$) relative to the moving observer. Thus:

$$\Xi(in).t_B 1/\gamma + \Xi(out).t_B 1/\gamma = 2t_B 1/\gamma$$

So:

$$\Xi(in) + \Xi(out) = 2$$

The previously calculated apparent rates for a stationary clock satisfy the above condition. QED: Apparent rates due to time delays vanish when the moving observer returns to the stationary clock. Thus, the only remaining effect seen by the moving observer on reunion is the anti-dilation of the stationary clock due to its absolute rest.

Conclusion

According to Machian Relativity, the overall effects of time delays vanish on reunion. Only the physical dilation changes persist. This applies to observers with absolute motion or at absolute rest. This conclusion must also apply to Special Relativity and Galilean Relativity. Therefore, when the moving twin (or clock) returns to the earth, time delays do not have a residual effect that hides the paradoxical symmetry of Special Relativity. In his original paper on Special Relativity (1905), Einstein did not address the asymmetry issue. In later years he favoured General Relativity as the explanation, rather than time delays.

And physicists do not use this method in practical applications of Special Relativity (e.g. the decay rates of continuously non-inertial relativistic subatomic particles in accelerators). Rather, they use the theory in an arbitrarily asymmetric way, and then claim experimental results "prove" Special Relativity. And if asked to explain this asymmetry (for continuous acceleration), they mention accelerations or General Relativity. But they never derive the acceleration asymmetries mathematically from First Principles. Also, a clock *can* be compared to a system of stationary synchronised clocks, while it moves inertially. In this case, the symmetry problem of Special Relativity cannot be circumvented.

The only reason why scientists and authors believe in the existence of reciprocal time dilation is because the equations of Special Relativity predict such a thing. It is only ever applied to the Twins Paradox, but *never to practical applications*, such as the Hafele-Keating Experiment. We know that two parallel currents of *inertially moving electric charges* are magnetically attracted, while those at rest on the earth are not. Thus, inertial motion <u>cannot</u> be relative per-se. Machian relativity gives the correct answer. Inertial and accelerated systems are different due to acceleration relative to the "Fixed Stars", and every inertial system has a physical time dilation due to motion relative to the "Fixed Stars". In this way <u>the Laws of Physics are the same</u>. And, *such asymmetric time dilation is more consistent with the asymmetric "gravitational dilation" of General Relativity.* Consider clocks in the *proper* gravitational field of a massive body (rather than the *bogus* induced fields of the Equivalence Principle). Relative to clocks at low gravitational potentials, clocks at higher potentials have *faster rates*, whereas relative to clocks in high gravitational potentials, clocks at lower potentials have *correspondingly slower rates*. This occurs irrespective of whether the clocks are stationary in the gravitational field or if they are in orbital free-fall. A clock in orbital free-fall is locally inertial and if its gravitational dilation remains asymmetric, there can be *no objection* to clocks undergoing asymmetric dilation during inertial motion. Similarly, we know from the Hafele-Keating Experiment that the velocity of an uncorrected clock orbiting the earth produces asymmetric time dilation. As the frame of the clock is locally inertial, it is likely that a similar clock moving inertially from the earth will also have asymmetric time dilation. This gives the *simplest resolution* of the Twins Paradox.

MU

MACH'S PRINCIPLE AND RELATIVISTIC LIGHT SOURCES

The previous section on time delay effects and observed time dilation allows an analysis of light sources and relativistic effects on brightness and frequency.

Consider a light source at absolute rest in the Plenum. Say the source emits energy (per unit area) E_0 over a duration T_0 relative to a stationary observer. For simplicity assume the emission is non-divergent. So brightness $W_0 = E_0 / T_0$. Now let the source have some absolute motion. All processes are time dilated, and this includes the emission of photons. Therefore, relative to stationary observers, the brightness of the moving source will tend to decrease, as the rate of photon emission is now stretched over a duration T_0 / γ. However, due to relativistic energy increase, the energy emitted by the source is now E_0 / γ, because each photon is emitted with correspondingly more energy. So relative to stationary observers we now have $W_\alpha = E_\alpha / T_\alpha = (E_0 / \gamma) / (T_0 / \gamma) = W_0$. Also, we must consider if the source is receding or approaching, as this will continuously change the time delays for the energy exchange relative to the static observer. This means the observed brightness is modified in the same way as is the time dilation of a moving clock relative to a static observer. So if the source is approaching, we have $W_\alpha = W_0 / (1 - v_\alpha/c)$, and if the source is receding, the observed brightness is $W_\alpha = W_0 / (1 + v_\alpha/c)$. For divergent (e.g. spherical emission), there is the added effect of brightness changing as the distance varies due to approach or recession. In this case, the average brightness $\{W_0\}$ can be used.

For observers moving with the source, they (and their clocks) are correspondingly time dilated, so they experience no relative change in the duration of energy emission. Thus $T_B = T_0$. Their own mass-energy (and everything else in the system) has also increased in the same proportion as the energy from the source, so they experience no relative change in its energy, so $E_B = E_0$. Thus they experience no relative change in the source brightness, $W_B = E_B / T_B = W_0$. However, if the source is instead at absolute rest, the changes relative to moving observers are inverted (this is because in Machian Relativity changes are asymmetric relative to distant matter). So, the relative duration of energy emission is γT_0, and the relative energy is γE_0. Thus relative to moving observers, the observed brightness of a source at absolute rest is $W_B = \gamma E_0 / \gamma T_0 = W_0$. By including the additional effects of time delays, we have $W_B = W_0(1 + v_\alpha/c)$ for an approaching source, and if the source is receding, the observed brightness is $W_B = W_0(1 - v_\alpha / c)$.

Machian Relativity and Relativistic Doppler Effects

In accordance with mass-energy increase, the energy of emitted photons increases as the absolute speed of the moving light source increases. Relative to stationary observers, this corresponds to a greater photon frequency given by $f_\alpha = E_0 / h\gamma = f_0 / \gamma$. As before due to changing time delays, the emitted light will be blue-shifted if approaching a stationary observer $f_\alpha = f_0 / \gamma(1 - v_\alpha/c)$, and red-shifted when receding $f_\alpha = f_0 / \gamma(1 + v_\alpha/c)$. This is because the motion of light is absolute relative to the Plenum (and thus to stationary observers). For other directions, the magnitude of the frequency shift is determined by the angle of light relative to the source's absolute motion.

Next we must consider how the frequency of light-waves varies relative to moving systems. Let a stationary source emit a pulse. The speed of the pulse is determined by the motion of a fixed point on the wave (say a peak or trough). Relative to a stationary system, the time for a peak to travel from the origin to a distance of one wavelength is equal to the period T_α of the wave. So its distance from the origin is cT_α. Relative to stationary observers, the distance travelled by an inertial system with absolute velocity v, moving in the same direction of the wave and in the same time, is

$v_\alpha T_\alpha$. So relative to stationary observers, the peak (or trough) travels to a point on the axis of the moving system whose distance is T_α $(c - v_\alpha)$ from the moving origin. When the origins of both systems initially coincided, a clock in the moving system at this point from its origin would read $-v_\alpha T_\alpha$ $(c - v_\alpha)/c^2\gamma$ due to relativistic simultaneity. Thus after a time T_α in the stationary system, moving clocks will advance by γT_α. Thus the corresponding reading on the moving clock (when the peak reaches it) is the period of the wave for the moving system. This is:

$$T_B = -v_\alpha T_\alpha (c - v_\alpha)/c^2\gamma + \gamma T_\alpha = T_\alpha \sqrt{(c - v_\alpha)} / \sqrt{(c + v_\alpha)}$$

So: $f_B = 1 / T_B = f_\alpha\sqrt{(c + v_\alpha)} / \sqrt{(c - v_\alpha)}$

This frequency change relative to moving observers is apparent, and arises from the time dilation and relativistic simultaneity effects within the system itself. This equation applies to any frequency of light, including those that have been modified due to emission from sources with relativistic absolute motion. This is because the wavefronts from sources always have absolute motion relative to the Plenum, but not to the source itself. Thus, if a source and observer move with the same absolute motion and, the observer is ahead in the direction of absolute motion, then:

$$f_B = f_0\sqrt{(c + v_\alpha)} / \{\gamma(1 - v_\alpha/c) \sqrt{(c - v_\alpha)}\}$$
$$= f_0 / (1 - v_\alpha/c)^2 \neq f_0$$

This is because in Machian Relativity sources and systems undergo objectively real changes. So we cannot adopt the position of Special Relativity, where if a system and source move with the same inertial velocity, then $f_B \equiv f_0$ because "there is no such thing as absolute motion, and so the term v_α has no meaning".

Corollary

The previous method measures the passage of a fixed point on a wave between different points in both systems. A better method is to consider a fixed point in either system, say the origins. Then, the passage of two points on the wave can be considered instead, say two peaks separated by a wavelength. This method is better than the previous one because simultaneity changes in the moving system can be factored out. Relative to the stationary system, one wavelength is cT_α, where T_α is the period. Let a peak be at the origins of both systems when they coincide. Relative to the stationary system, the next peak is one wavelength away at $-cT_\alpha$. Relative to stationary observers, the position of this peak is $-cT_\alpha + ct_\alpha$, and the position of the moving system is $v_\alpha t_\alpha$. Thus when this peak reaches the origin of the moving system, we have $v_\alpha t_\alpha = ct_\alpha - cT_\alpha$. So $t_\alpha = T_\alpha c / (c - v_\alpha)$. Because of time dilation in S_B, this duration between the arrival of two peaks is $t_B = \gamma\, t_\alpha = \gamma T_\alpha c / (c - v_\alpha)$. This is the period of the wave relative to the moving system, $t_B = T_B$. Thus:

$$T_B = \gamma T_\alpha c / (c - v_\alpha)$$

Now, the frequency of the wave is the inverse of its period, thus:

$$f_B = f_\alpha (c - v_\alpha) / c. \ \gamma = f_\alpha (1 - v_\alpha/c) / \gamma$$

This is essentially the same result as for apparent dilation effects. Observers with absolute motion measure the frequency to be changed due to an anti-dilation factor $1 / \gamma$, and an additional recession factor $(1 - v_\alpha/c)$ due to changing time delays, which becomes $(1 + v_\alpha/c)$ if moving observers approach the stationary source. So if a source has the same velocity as the moving system, stationary observers measure the frequency in the direction of the source's motion to be $f_\alpha = f_0 / \gamma(1 - v_\alpha/c)$. That is, blue shifted. Therefore, observers moving with such a source determine this frequency to be:

$$f_B = f_\alpha (1 - v_\alpha/c) / \gamma = f_0 / \gamma^2 \neq f_0$$

Hence observers in the inertial system S_B do not measure a relatively stationary source to have a frequency $f_B = f_0$. This is because both the system and the source are physically changed by their absolute motion according to Mach's Principle. These calculations show that an observer with a high velocity (relative to the external Universe) will see approaching stars to be brighter and blue shifted, while receding stars will appear darker and red-shifted.

Corollary

The spherical wavefronts of both the stationary and moving sources remain centred on points of absolute rest. So an observer moving with a source with absolute motion does not remain centred on its source's wavefronts. And because of Lorentz contraction of moving measuring rods in the direction of motion, the wavefronts are no longer spherical for a moving observer. But, length contraction, time dilation and simultaneity changes ensure the speed of light is constant for a moving observer. Also, because of the absolute motion of light, the paths of light rays emitted from a source with absolute motion will vary relative to its observers, depending on the source's absolute velocity (like stellar aberration). Now, this, along with other asymmetries in Machian Relativity, contradicts conventional science, which claims there is no absolute motion. But such asymmetries must still arise during acceleration, so they cannot be avoided. What I am saying is that it is conceptually simpler for asymmetries to occur during both acceleration and inertial motion (after the acceleration has stopped). And the cause is the influence of distant matter (the external universe), which continues to act according to Mach's Principle.

The Human Eye

Consider the human eye to approximate to a pin-hole camera. When an observer looks in the direction of his absolute motion, his eyes are Lorentz contracted (along their axis). The size of an image on the retina is therefore reduced, and so objects appear smaller. When looking sideways to his motion, his eyes are still contracted in the direction of their absolute motion (which is now transverse to the axis of the eye). This means images have a larger size on the retina (in the direction of the absolute motion), and so objects appear extended in the direction of this motion.

NU

VARIOUS REFUTATIONS OF EINSTEINIAN ELECTRODYNAMICS

The tragedy of science is when a beautiful theory is ruined – by an ugly fact. (T. Huxley)

The Magnetism Paradox

It is not recognised that Einstein's theory of Special Relativity – as usually understood at the present time – when applied to Maxwell's Electrodynamics, leads to inconsistencies which *cannot possibly be inherent in any natural phenomena*. Take, for example, the supposed reciprocal electrodynamic action between two identical electrically charged systems moving relatively to each other at inertial speed v. In system A (S_A) let there be two charged bodies, at rest and perpendicular to the relative motion. There are identical bodies in system B (S_B). All the charged bodies have the same shape, which can be (for example) a sphere or rod of any arbitrary length. According to Special Relativity, there is complete symmetry between the two systems. That is, systems A and B are completely equivalent to each other, so that; relative to A things change according to their speed relative to A, and relative to B things change according to their speed relative to B.

Thus, relative to A (e.g. relative to an observer at rest in A), there is no magnetic attraction between A's charges, while at the same time there is a magnetic attraction between B's charges. And if the relative speed is sufficiently great, the magnetic attraction will overcome the electrostatic repulsion, causing the two charged bodies in B to be mutually attracted and eventually touch – an objective fact which must be true for observers in both A and B. An example of such behaviour is the pinch effect used to initiate nuclear fusion reactions in heavy hydrogen plasma.

But according to Einstein's Relativity Postulate, relative to B, its charges are at rest so there will be no magnetic attraction between them, while A's charges are instead moving – generating a magnetic attraction that eventually causes them to touch. This too is an objective fact which must be true for both observers in A and B. Therefore, for observers in both systems, charges in A are simultaneously touching and not touching. And similarly for all observers, charges in B are simultaneously touching and not touching. Thus, according to the Relativity Postulate, unphysical and contradictory results are generated in both systems. This is a paradox which I call the *Paradox of Einsteinian Magnetism*.

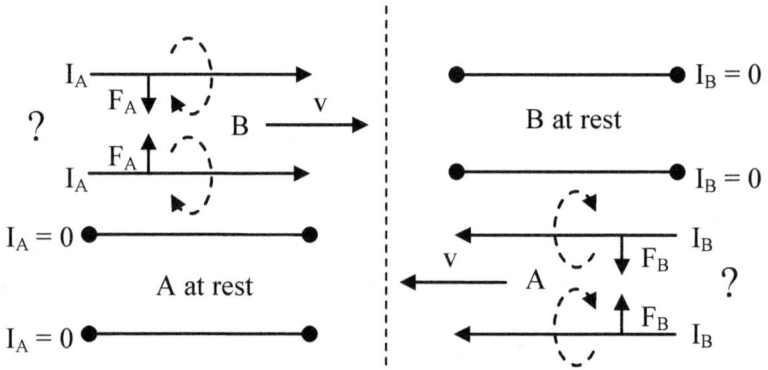

Fig.39: The Einsteinian Magnetism Paradox

Examples of this sort, together with the impossibility of finding any experimental evidence for symmetric time dilation (as predicted for inertial systems by Special Relativity), suggest that the phenomena of electrodynamics as well as of mechanics possess no properties corresponding to Einstein's idea of relative rest.

NU I
The Einsteinian Paradigm

Einstein assumes covariance for Maxwell's Equations, and solves the equations accordingly. This is his attempt to bring electromagnetism in line with his Special Relativity, but it is not a proof as such that electromagnetism is inherently covariant. He considers three scenarios in an inertial system; induction in a coil (such as by a nearby electromagnet or moving bar magnet), displacement current (magnetic curl induction where the coil is charged), and magnetism due to an electric current. The corresponding Maxwell equations are:

$$\text{curl } E = -(1/c).\partial H/\partial t$$
$$\text{curl } H = (1/c).\partial E/\partial t$$
$$\text{curl } H = (1/c).\rho v$$

He then determines how these equations transform to a second system moving inertially relative to the first using the Lorentz Equations. The first two equations produce a set of relativistic field equations describing how electric and magnetic fields are inter–related between both systems. For the last equation, the Magnetic Paradox occurs, because his Relativity Postulate requires that observers moving with an electric current should experience no magnetic field for the current, which is clearly erroneous.

Authors of science books then make specious claims that Special Relativity "transforms away" the magnetic fields of moving charges. But this has never been proved experimentally. It is quite clear however, that the effects of electromagnetism do not "transform away". Two wires carrying the same polarity current are attracted and eventually touch. This fact of touching must be objectively true for both external observers and for observers in the same frame of the moving charges in the current.

The same reasoning applies to the pinch effect during nuclear fusion. The heavy hydrogen nuclei circulating in a tokomak (toroidal reactor) constitute a current and they are attracted, overcoming their electric repulsion. Eventually the separation is sufficiently reduced, generating nuclear fusion reactions. These reactions are an objective fact that applies to any observer. And if we conduct a "thought experiment", this fact will also apply to observers moving with the nuclei. So, the frame of the moving charges cannot be considered as being "at rest". And, by extension to currents in straight wires, this effect must also occur if the nuclei move inertially in a straight line (while Einstein's theory gives the false result that the inertial nuclei are "at rest" and so should not undergo pinch effects). Therefore, Einstein's concept of inertial equivalence is refuted.

Electromagnetic Induction and Relative Motion

When a conducting coil and magnet are moved relative to each other, an electric current is induced in the coil. According to Maxwell's electromagnetic theory, such effects were transmitted by the Luminiferous Aether. This meant that although the outcome was the same, the process by which induction occurred differed according to whether the magnet or coil moved relative to the Aether. According to Einstein, this asymmetry was "intolerable". He reasoned that there was no Aether, and instead induction depended solely on the relative motion between the magnet and coil, regardless of how they were caused to move. This meant there was now one explanation, and the asymmetry had been removed. To Einstein, this was another example of "everything is relative".

However, the analysis needs clarifying. The underlying cause of electromagnetic induction is that a conductor is subjected to a *changing* magnetic field, which means there need not be any relative motion. For example, if a coil is instead situated near a second coil carrying a current, the second coil is an electromagnet (in contrast to a permanent magnet). If the current remains constant, the electromagnet generates a constant magnetic field, and no current is induced in the first coil. If however, the current in the electromagnet is changed, its magnetic field changes, and this induces a current in the first coil. This occurs *without any relative motion between both coils.* A further consequence of this is that if a conductor moves relative to any magnetic source which has a spatially uniform field, then *no current* will be induced in the conductor. For example, let two magnets be placed near each other, so that the north pole of one is near the south pole of the other. Therefore, between the magnets an approximately uniform field occurs, while around the further ends of the magnets the field decreases through space. If a coil moves relative to either of the outer magnetic poles, the non–uniform field induces a current in the coil. If, instead, the coil moves in the same way, but now between both magnets, the uniform field does not induce a current in the coil. So, Einstein's analysis of effects depending on relative motion does not apply to this scenario. So generally, the effect depends on relative *interactions* between a conductor and magnetic field.

NU II
The Einstein–Lorentz Relativistic Field Equations

Consider Maxwell's equations for current induction in a coil and displacement current (rotational magnetic field). These are, respectively:

$$(1/c).\partial H/\partial t = -\text{curl } E$$
$$(1/c).\partial E/\partial t = \text{curl } H$$

Relative to observers in A, let an electric field $E(x, y, z)$ have vectors (E1, E2, E3), and a magnetic field $H(x, y, z)$ have vectors (H1, H2, H3). In order to simplify Einstein's method, consider a magnet and charged coil whose axes are aligned with the x–axis of A (allowing some terms to vanish). Therefore:

(1a) $(1/c).\partial H1/\partial t_A = \partial/\partial z_A[E2] - \partial/\partial y_A[E3]$
(1b) $(1/c).\partial E1/\partial t_A = \partial/\partial y_A[H3] - \partial/\partial z_A[H2]$

The full set of generalised equations is described in relativity textbooks, and will not be dealt with here. The Lorentz Equations are used to transform all the equations to a second system B moving at inertial speed v relative to A. For observers in B, the above equations become:

(2a) $(1/c).\partial H1/\partial t_B = (1/\gamma).\partial/\partial z_B [E2 - H3.v/c] -$
$(1/\gamma).\partial/\partial y_B [E3 + H2.v/c]$
(2b) $(1/c).\partial E1/\partial t_B = (1/\gamma).\partial/\partial y_B [H3 - E2.v/c] -$
$(1/\gamma).\partial/\partial z_B [H2 + E3.v/c]$

In other words, the equations for observers in B are modified because the space–time of B changes due to motion relative to A.

Now, in order to satisfy Einstein's Relativity Postulate, observers in B must consider themselves "at rest". Therefore, their electromagnetic equations must be covariant with those for A's observers. Thus:

(3a) $(1/c).\partial H1'/\partial t_B = \partial/\partial z_B\,[E2'] - \partial/\partial y_B\,[E3']$

(3b) $(1/c).\partial E1'/\partial t_B = \partial/\partial y_B\,[H3'] - \partial/\partial z_B\,[H2']$

Where (E1', E2', E3') and (H1', H2', H3') are the respective force vectors for fields E and H in system B. Einstein then used equations (2a, 2b) with (3a, 3b) to find the relativistic field equations linking (E1, E2, E3) and (H1, H2, H3) with (E1', E2', E3') and (H1', H2', H3'). These are:

$$E1' = E1 \qquad\qquad H1' = H1$$
$$E2' = [E2 - v.H3/c]\,/\gamma \qquad H2' = [H2 + v.E3/c]\,/\gamma$$
$$E3' = [E3 + v.H2/c]\,/\gamma \qquad H3' = [H3 - v.E2/c]\,/\gamma$$

These equations were derived by Einstein and are identical to those derived by Lorentz.

Corollary I

Algebraic rearrangement of the above gives the following:

$$E1 = E1' \qquad\qquad H1 = H1'$$
$$E2 = [E2' + v.H3'/c]\,/\gamma \qquad H2 = [H2' - v.E3'/c]\,/\gamma$$
$$E3 = [E3' - v.H2'/c]\,/\gamma \qquad H3 = [H3' + v.E2'/c]\,/\gamma$$

These equations are covariant with the previous set. This is unlike (say) time dilation, where algebraic rearrangement the equation is asymmetric. For example, if $t_B = \gamma.t_A$, then algebraic rearrangement gives $t_A = t_B\,/\gamma$. This is because the above equations are based on two quantities (e.g. E and H), whereas the space-time variables occur alone (e.g. t, or x).

For Special Relativity, the situation is ambiguous. Of course, scientists will say, "Ah, but Special Relativity is covariant, and that in fact $t_A = \gamma.t_B$". But this could be because B is "at rest", or it could be apparent covariance due to simultaneity (as occurs for Machian Relativity).

Corollary II: Electromagnetic Covariance

In fact, the equivalent process of transforming these equations from B to A using the Lorentz Transform for B can be undertaken. Because of the Relativity Postulate, B can be assumed to be "at rest", with A's space–time changing in a reciprocal way relative to B. For B "at rest" we have:

(4a) $(1/c).\partial H1'/\partial t_B = \partial/\partial z_B[E2'] - \partial/\partial y_B[E3']$

(4b) $(1/c).\partial E1'/\partial t_B = \partial/\partial y_B[H3'] - \partial/\partial z_B[H2']$

The Lorentz Transforms for transforming coordinates from B to A gives:

(5a) $(1/c).\partial H1'/\partial t_A = (1/\gamma).\partial/\partial z_A [E2' + H3'.v/c] -$
$(1/\gamma).\partial/\partial y_A [E3' - H2'.v/c]$

(5b) $(1/c).\partial E1'/\partial t_A = (1/\gamma).\partial/\partial y_A [H3' + E2'.v/c] -$
$(1/\gamma).\partial/\partial z_A [H2' - E3'.v/c]$

We require the equation in A to have the same form as in B. So:

(6a) $(1/c).\partial H1/\partial t_A = \partial/\partial z_A [E2] - \partial/\partial y_A [E3]$

(6b) $(1/c).\partial E1/\partial t_A = \partial/\partial y_A [H3] - \partial/\partial z_A [H2]$

Thus:

$$E1 = E1' \qquad\qquad H1 = H1'$$
$$E2 = [E2' + v.H3'/c]/\gamma \qquad H2 = [H2' - v.E3'/c]/\gamma$$
$$E3 = [E3' - v.H2'/c]/\gamma \qquad H3 = [H3' + v.E2'/c]/\gamma$$

Therefore, the equations for A's observers describing induction are covariant with B's equations. Lorentz did not include this step. Einstein did not include this step either, although the covariance of his theory allowed it.

Corollary III: The Transformation Paradox

Let a circular coil in A be subjected to a time varying magnetic field (H1, 0, 0). This generates electric curl in the coil (0, E2, E3). The corresponding curl in B is:

$$(0, E2', E3') = (0, E2/\gamma, E3/\gamma)$$

Transforming back to A we have:

$$(0, E2, E3) = (0, E2'/\gamma, E3'/\gamma) = (0, E2/\gamma^2, E3/\gamma^2)$$

Thus there is a paradox, similar to the time dilation paradox. These equations violate the rules of algebra and physical change. According to algebra and physical change, we would expect:

$$(0, E2, E3) = (0, \gamma E2', \gamma E3')$$

A similar paradox occurs for magnetic curl also:

$$(0, H2, H3) = (0, H2'/\gamma, H3'/\gamma) = (0, H2/\gamma^2, H3/\gamma^2)$$

It will be argued that the equations of Special Relativity are "kinematic" and "unphysical", but this is false. Also, like for time dilation, the asymmetric form above must occur when B is accelerated by applied forces.

Corollary IV

For observers in system B, their measurements of the coil–magnet interaction in A require the use of equations (2a, 2b), which have a different form from those of A (1a, 2b). So, observers in system B must deduce their clocks and rulers have changed, due to motion relative to A. [Because, if system B had not changed, then Galilean relativity would still be applicable, and its equations would have the same form as for observers in A]. On this basis, system B cannot be regarded as being "at rest".

Glyn Phillips

Corollary V: The Magnetism Paradox

The same reasoning is also applied to electric ("convection") currents. In system A let there be a current given by vectors (J1, J2, J3) such that:

$$J_A = (J1, J2, J3)$$
$$= (u1\rho, u2\rho, u3\rho)$$

Where (u1, u2, u3) is the charge velocity vector and ρ the charge density. From Maxwell's Equations, the magnetic field associated with this current is given by:

$$(1/c).J1 = (1/c).u1\rho = \partial/\partial y_A[H3] - \partial/\partial z_A[H2]$$
$$(1/c).J2 = (1/c).u2\rho = \partial/\partial z_A[H1] - \partial/\partial x_A[H3]$$
$$(1/c).J3 = (1/c).u3\rho = \partial/\partial x_A[H2] - \partial/\partial y_A[H1]$$

Again, because of Einstein's Relativity Postulate, system B is regarded as at rest, so its equations are covariant with those of system A. Therefore, when observers in B measure the properties of the current in A, we must have:

$$(1/c).J1' = (1/c).u1'\rho' = \partial/\partial y_B[H3'] - \partial/\partial z_B[H2']$$
$$(1/c).J2' = (1/c).u2'\rho' = \partial/\partial z_B[H1'] - \partial/\partial x_B[H3']$$
$$(1/c).J3' = (1/c).u3'\rho' = \partial/\partial x_B[H2'] - \partial/\partial y_B[H1']$$

Because of B's space–time changes due to its motion v relative to A, the corresponding charge velocity relative to B is given by the usual velocity transform equations:

$$u1' = (u1 - v) / [1 - u1.v/c^2]$$
$$u2' = \gamma.u2 / [1 - u1.v/c^2]$$
$$u3' = \gamma.u3 / [1 - u1.v/c^2]$$

Similarly, the charge density of this current relative to B is:

$$\rho' = (1/\gamma).[1 - u1.v/c^2].\rho$$

For example, consider a current flowing along the x–axis of A,

151

such that $u = (u1, 0, 0)$. If B moves at the same velocity as the current velocity, then $v = u1$, which gives $u1' = 0$. Therefore:

$$(1/c).J1' = (1/c).u1'\rho' = \partial/\partial y_B[H3'] - \partial/\partial z_B[H2'] = 0$$

This means that while the charges have a magnetic field in system A because they are moving, relative to B they are not moving and so have no magnetic field. This satisfies Einstein's Relativity Postulate, where there is no absolute motion and any inertial system can be considered as being at rest. Initially, this appears plausible, but closer consideration reveals the following paradox: For two parallel currents in A, the charges are magnetically attracted, and this must be objectively true for observers in both A and for those in B. But, according to Einstein's Relativity Postulate, the same charges are at rest relative to observers in B and so should not be attracted for these observers. So there is a contradiction.

Corollary VI

For Einstein's treatment of electric current, the Magnetism Paradox has been asserted (and resolved by my reasoning). In addition, his method also contains the Velocity Paradox. If observers (say) in B measure a current to have velocity $u1'$ according to the relativistic velocity equations, this happens because system B has physically changed due to motion relative to A (thereby refuting the erroneous idea that the equations of Special Relativity are "kinematic"). This follows, because if B did not physically change, then observers in B would determine $u1'$ according to Galilean Relativity, where $u1' = u1 - v$.

Now, it can be seen that this methodology mixes ideas on states of motion: the velocity transform equation for the current relative to B assumes B has physically changed (so its observers determine they are not at rest, thereby contradicting the Relativity Postulate), while the equation for a current's magnetism assumes it arises due to charge velocity relative to B (so its observers determine they are at rest, thereby agreeing with the Relativity Postulate). In my method (Machian Relativity), B's observers determine they have absolute motion by both processes (current velocity and magnetism).

Also, for Galilean Relativity, we have $u1' = u1 - v$, and $u1 = u1' + v$. But according to Special Relativity, if A measures $u1$ and v, then B instead measures a modified (relativistic) velocity $u1'(r) = (u1 - v) / [1 - u1.v/c^2]$. Thus observers in B measure $u1'(r) \neq u1'$ because their space-time has changed. Likewise $u1 = (u1'(r) + v) / [1 + u1'(r).v/c^2]$.

But, according to the Relativity Postulate[9], system B is also "at rest" (for its observers), and instead that A's space-time should change due to its motion relative to B. In this case, observers in B measure the current velocity to be $u1'$, not the relativistic $u1'(r)$. Instead, observers in A measure the current to have a relativistic value $u1(r) \neq u1 = (u1' + v) / [1 + u1'.v/c^2]$. So: $u1(r) \neq u1 \neq u1' + v$. Thus there is a paradox. Like in the Clock (Twins) Paradox of Time Dilation, acceleration cannot be used as an excuse as measurements can be performed in the absence of acceleration (i.e. while the two systems are in relative inertial motion).

---

[9] If two relatively moving systems are initially considered in isolation from other matter, the more massive system must have a greater effect on the less massive one than vice versa. Hence their space-time changes cannot be reciprocal. And Special Relativity's dilation asymmetry in the Hafele-Keating Experiment must still apply if both the earth and satellite clocks are instead on satellites, relatively moving in different orbits. Such clocks are in free-fall, so must be regarded as inertial. Thus it is likely that a similar asymmetry will occur if one (or both) of the clocks has uniform rectilinear motion relative to the earth, and so Einstein's Relativity Postulate must be false. There is no specific experimental evidence for reciprocal time dilation. The concept has no practical application and is only ever used in the Twins Paradox, where its paradoxical consequences need rectifying with additional phenomena. Hence, scientists' use of this concept is pseudoscience.

XI

MACH'S PRINCIPLE AND ELECTROMAGNETISM:
Magnetism by motion relative to Matter

The Magnetism Paradox is resolved by abandoning Einstein's concepts of symmetry and using the combined average method described previously for clocks and rigid rods in Mach fields.

Each system is surrounded by its own Mach field, arising from its own mass. Therefore A *moves through* the Mach field of B, and B *moves through* the Mach field of A. So electric charges in A have a magnetic field determined by their motion relative to B; *regardless of A and B's spatial separation*. But these charges are also at rest relative to their own Mach field (mass), and this counters the effect of motion relative to B (Self-influence Principle). Therefore, the resultant magnetic field around the charges is determined by motion relative to the centre of gravity (average momentum) of A and B. For example, if both systems have equal masses, then the magnetic field is determined by $v/2$. The same reasoning applies to B's charges moving relative to A. Even if there are no charges present in one system, charges in the other system still have magnetic fields determined by the same reasoning (because the effect is due to charges moving relative to the Mach fields of other masses, not relative to other charges).

In other words, if a charge tends to have no magnetic field due to being at rest relative to one observer, and a magnetic field due to motion relative to another observer, then for both observers the charge has an average (partial) magnetic field.

If a third system C is included (which may or may not have electric charges), then the magnetism of A's and B's charges is determined by motion relative to the combined average momentum of A, B and C. And from this, it follows that *all other masses in the universe* must be included, as they also have Mach fields acting across space. As a result, <u>the magnetic field of an electric current arises due to motion relative to the Plenum (average momentum of all matter)</u>. So, according to the Theory of Machian Relativity, in Maxwell's equations, the velocity v is relative to the average momentum of all matter and is *absolute*. Contrary to Einstein's assertions, *there is no reciprocal (relative) magnetic effect between systems*:

Machian Plenum

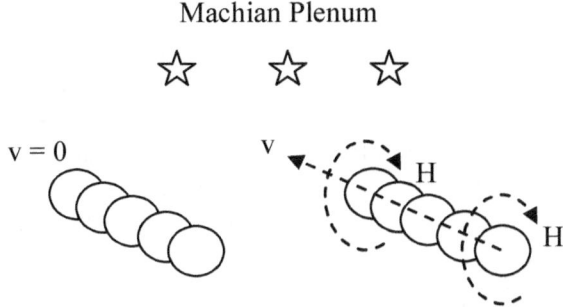

<u>Fig.40: Paradox-free Magnetism according to Mach's Principle</u>

The introduction of the Aether is unnecessary. Mach fields are already inherent in Einstein's physics to generate the reciprocal relativistic effects he claims to occur. But, the mass of each system resists the changes induced by the other, causing an average effect. These Mach fields allow the effects of relative mass to be included with relative velocity, once Einstein's Relativity Postulate has been abandoned. This means the reality of interacting Mach fields is realised. The outcome is that *relativistic asymmetries and absolute motion are self–generating from the interaction of all matter in the universe*. This resolves the paradox and unites electromagnetism with Mach's Principle. Thus, there is no need to "transform away" the magnetic field. An observer moving with an electric current experiences the resulting magnetic field, and makes the same deduction as stationary observers that his system has absolute motion relative to the Plenum (external universe). Maxwell's equations are therefore *invariant* for any system, and the Laws of Physics are the same for different observers. So, the effects of relative mass are included with the effects of relative velocity. Mach fields are the <u>causal mechanism</u> that explains how systems affect each other across space. If their existence is denied, then it is <u>causally impossible</u> for a system to make its presence felt in the vicinity of another to generate such magnetic effects.

In addition, there is no need to calculate the individual momentum of all masses in the universe to determine an electric current's motion relative to the Plenum. More massive bodies are harder to accelerate that less massive ones, so celestial bodies (e.g. the Fixed Stars) are a reasonable approximation to absolute rest. For example, the earth approximates to a system of absolute rest when studying relativistic particles. In other cases, the absolute motion of the earth may be significant, in which case the sun will be a better approximation to a stationary system. Thus, for electric currents, the frame of the earth is a reasonable approximation to absolute rest.

<u>Consequences of Machian Electromagnetism:
Influence of External Universe on Systems of Electric Charges in
Relative Motion</u>

Consider two systems in relative motion to each other. According to Machian Relativity, the effects of the external universe on each system need to be included. Therefore, let system A be at rest in the Plenum, and let system B have an absolute velocity, due to its inertial motion relative to the external universe. The frame of the earth can be considered to be an approximation to system A. In each system there are two electric charges arranged perpendicular to the relative motion. Using their clocks and rulers, observers in A determine their charges move according to the inverse square law of electric repulsion. However, they determine that the interaction of B's charges is affected by electric repulsion, magnetic attraction and mass increase. For B's observers, their measurement of the interaction between A's charges is a result of their mutual electric repulsion and B's relativistic space–time changes. B's observers also measure the interaction between their own charges and conclude it results from electric repulsion, magnetic attraction, mass increase and the changes in their own clocks and rulers. If the magnetic attraction between B's charges is greater than their electric repulsion, B's time dilation cannot transform the net attraction back to a normal repulsion for B's observers. This means that Einstein's Relativity Postulate must be false. Even if this is denied, this electromagnetic asymmetry must still occur if B is accelerated by applied forces relative to A (or vice versa). However, Machian Relativity properly derives the asymmetry from first principles, whereas in Special Relativity the asymmetry is merely assumed from symmetric equations. So, Machian Relativity eliminates any paradoxes *at source, in the equations themselves*.

If the charges in B are arranged parallel to their motion, each charge still has its associated magnetic field, but the field component between the charge pair is zero. So, A's observers determine that B's charges move instead according to their electric repulsion and their mass increase. In addition, we may conjecture an additional influence on how B's charges interact − that their electric fields are Lorentz contracted in the direction of their absolute motion [This follows because inertial bodies are Lorentz contracted, and the absence of applied forces means such bodies are not under mechanical compression. Therefore, the cause must be that the force fields holding the atoms of the body in equilibrium are also Lorentz contracted]. The effects of the external universe on induction, if systems A and B each have their own coil and magnet, will be considered later.

OMICRON

MACHIAN ELECTRODYNAMICS

Transforming Maxwell's Equations according to Mach's Principle

According to the Theory of Machian Relativity, let system A be at rest relative to the external universe, and let system B have absolute velocity v_p (relative to the external universe). If a magnet and charged coil interact, relative to A we have:

$$(1/c).\partial H1/\partial t_A = \partial/\partial z_A[E2] - \partial/\partial y_A[E3]$$
$$(1/c).\partial E1/\partial t_A = \partial/\partial y_A[H3] - \partial/\partial z_A[H2]$$

These are the respective equations for electromagnetic induction and displacement current. The Mach-Lorentz Transform to system B gives:

$$(1/c).\partial H1/\partial t_B = (1/\gamma).\partial/\partial z_B [E2 - H3.v_p/c] -$$
$$(1/\gamma).\partial/\partial y_B [E3 + H2.v_p/c]$$
$$(1/c).\partial E1/\partial t_B = (1/\gamma).\partial/\partial y_B [H3 - E2.v_p/c] -$$
$$(1/\gamma).\partial/\partial z_B [H2 + E3.v_p/c]$$

Where $\gamma = \sqrt{(1 - v_p^2/c^2)}$. Transforming from A's field values to B's values, we can postulate equations for B covariant with those for A:

$$(1/c).\partial H1'/\partial t_B = \partial/\partial z_B [E2'] - \partial/\partial y_B [E3']$$
$$(1/c).\partial E1'/\partial t_B = \partial/\partial y_B [H3'] - \partial/\partial z_B [H2']$$

From this, the relation between the fields of A and B can be deduced. However, in Machian Relativity, this covariance does not imply that system B is somehow "at rest". The full set of equations is given as:

$$E1' = E1 \qquad\qquad H1' = H1$$
$$E2' = [E2 - v_p.H3/c] /\gamma \qquad H2' = [H2 + v_p.E3/c] /\gamma$$
$$E3' = [E3 + v_p.H2/c] /\gamma \qquad H3' = [H3 - v_p.E2/c] /\gamma$$

System B measures these fields, but unlike in Special Relativity, they are apparent, and arise from B's absolute space-time changes. From this, the apparent aspects can be factored out (by observers in B using these equations) to give the true field values (independent of B's changes). For example, these equations can be solved by considering a magnet and charged coil whose axes are aligned with the x-axis of system A.

Corollary I – Apparent Electromagnetic Covariance

Now, for system B we have:

$$(1/c).\partial H1'/\partial t_B = \partial/\partial z_B [E2'] - \partial/\partial y_B [E3']$$
$$(1/c).\partial E1'/\partial t_B = \partial/\partial y_B [H3'] - \partial/\partial z_B [H2']$$

Using the corresponding Mach-Lorentz Transform to change from B's to A's coordinates, we have:

(5a) $(1/c).\partial H1'/\partial t_A = (1/\gamma).\partial/\partial z_A [E2' + H3'.v_p/c] -$
$\qquad\qquad (1/\gamma).\partial/\partial y_A [E3' - H2'.v_p/c]$

(5b) $(1/c).\partial E1'/\partial t_A = (1/\gamma).\partial/\partial y_A [H3' + E2'.v_p/c] -$
$\qquad\qquad (1/\gamma).\partial/\partial z_A [H2' - E3'.v_p/c]$

We require the equation in A to have the same form as in B. So:

(6a) $(1/c).\partial H1/\partial t_A = \partial/\partial z_A [E2] - \partial/\partial y_A [E3]$
(6b) $(1/c).\partial E1/\partial t_A = \partial/\partial y_A [H3] - \partial/\partial z_A [H2]$

Thus:

$$E1 = E1' \qquad\qquad H1 = H1'$$
$$E2 = [E2' + v_p.H3'/c] /\gamma \qquad H2 = [H2' - v_p.E3'/c] /\gamma$$
$$E3 = [E3' - v_p.H2'/c] /\gamma \qquad H3 = [H3' + v_p.E2'/c] /\gamma$$

Corollary II – Factoring out Apparent Electromagnetic Covariance

Let a circular coil in A be subjected to a time varying magnetic field (H1, 0, 0). This generates electric curl in the coil (0, E2, E3). The corresponding curl in B is:

$$(0, E2', E3') = (0, E2/\gamma, E3/\gamma)$$

Transforming back to A we have:

$$(0, E2, E3) = (0, E2'/\gamma, E3'/\gamma) = (0, E2/\gamma^2, E3/\gamma^2)$$

Thus, there appears to be a paradox; the covariance of the equations is violating the rules of algebra and generating an inconsistent result. However, in Machian Relativity, B's space-time is physically changed by its absolute motion. Its covariance is *apparent and arises from its simultaneity changes.* And like the results for time dilation and length contraction, when the simultaneity changes are factored out (Appendix VI), the gamma factor is inverted ($\gamma \rightarrow 1 / \gamma$). So the covariance becomes an invariance, and the true difference in A's measurements relative to B's can be found. Therefore:

$$(0, E2, E3) = (0, \gamma E2', \gamma E3') = (0, E2, E3)$$

Corollary III

The same transformation applies to Displacement Current. If a magnet is subjected to a time varying electric field (E1, 0, 0), this generates magnetic curl (0, H2, H3) around the magnet. The corresponding curl in B is:

$$(0, H2', H3') = (0, H2/\gamma, H3/\gamma)$$

Transforming back to A we have:

$$(0, H2, H3) = (0, \gamma H2', \gamma H3') = (0, H2, H3)$$

Electric Currents and Machian Relativity

The same reasoning applies to electric currents. The magnetic field arises due to the absolute motion of the charges relative to the external universe. If two parallel currents are attracted and touch, this is an objective fact that is true for any observers, regardless of whether they are at absolute rest or "at rest" relative to the moving charges. For the rest system A, and moving system B, the equation is:

$$(1/c).J1 = (1/c).u1\rho = \partial/\partial y_A[H3] - \partial/\partial z_A[H2]$$
$$(1/c).J2 = (1/c).u2\rho = \partial/\partial z_A[H1] - \partial/\partial x_A[H3]$$
$$(1/c).J3 = (1/c).u3\rho = \partial/\partial x_A[H2] - \partial/\partial y_A[H1]$$

The resulting magnetic field is always due to the absolute motion of the current (u1, u2, u3) relative to the external universe (corresponding to system A), and is not due to motion relative to B. If system B has absolute motion v_p, the velocity transformation equations in Machian Relativity are:

$$u1' = (u1 - v_p) / [1 - u1.v_p/c^2]$$
$$u2' = \quad \gamma.u2 / [1 - u1.v_p/c^2]$$
$$u3' = \quad \gamma.u3 / [1 - u1.v_p/c^2]$$

However, these play no role in determining the current for observers in B. According to Mach's Principle, the current in A is $J1 = u1\rho$, where u1 is the absolute velocity of the current. If $u1 = v_p$ the Maxwell-Hertz equations for A are:

$$(1/c).\partial H1/\partial t_A \quad\quad = \partial/\partial z_A[E2] - \partial/\partial y_A[E3]$$
$$(1/c).\{\partial E1/\partial t_A + v_p\rho\} = \partial/\partial y_A[H3] - \partial/\partial z_A[H2]$$

Now, the situation in Special Relativity, where current transforms according to $J1' = u1'\rho'$ must be false, due to the Magnetism Paradox. Instead, let the corresponding current term for B be $v_p\rho'$. This must be so, as a current's magnetism in Machian Relativity is determined by its absolute motion relative to the Fixed Stars, and not by relative motion. Thus the corresponding electrodynamic equations for B are:

$$(1/c).\partial H1'/\partial t_B \qquad = \partial/\partial z_B [E2'] - \partial/\partial y_B [E3']$$
$$(1/c).\{\partial E1'/\partial t_B + v_p\rho'\} = \partial/\partial y_B [H3'] - \partial/\partial z_B [H2']$$

In the absence of a coil and magnet, this gives the current equation:

$$(1/c).v_p\rho' = \partial/\partial y_B [H3'] - \partial/\partial z_B [H2']$$

Using the previously derived equations, this gives:

$$(1/c).v_p\rho' = (1/\gamma).\partial/\partial y_B [H3 - E2.v_p/c] -$$
$$(1/\gamma).\partial/\partial z_B [H2 + E3.v_p/c]$$

For a constant electric current E2 = E3 = 0, so:

$$(1/c).v_p\rho' = (1/\gamma).\partial/\partial y_B [H3] - (1/\gamma).\partial/\partial z_B [H2]$$
$$= (1/\gamma).\partial/\partial y_A [H3] - (1/\gamma).\partial/\partial z_A [H2]$$
$$= (1/\gamma).(1/c).v_p\rho$$

[Note: $\partial/\partial y_B [H3] = \partial/\partial y_A [H3]$ and $\partial/\partial z_B [H2] = \partial/\partial z_A [H2]$, because the (y, z) axes are not affected by length contraction, and the magnetic curl is not (according to MR) dependent on motion relative to B].

Thus: $\qquad \rho' = \rho/\gamma$

Where, ρ' is the uncorrected charge density of the current, as measured by moving observers in B. According to MR, such observers measure the length contraction of stationary bodies as an apparent effect of simultaneity. So, if simultaneity effects in B are factored out, the true charge density for B is:

$$\rho'(true) = \gamma.\rho$$

This result is simpler than that for Special Relativity, which instead has:

$$\rho' = (1/\gamma).[1 - u1.v_p/c^2].\rho$$

Therefore in Machian Relativity, observers in B determine charges at relative rest to have non-zero magnetic fields given by:

$$H3' = H3/\gamma \text{ and } H2' = H2/\gamma$$

This is consistent with observers moving with two current carrying wires and measuring the subsequent magnetic attraction with their time dilated clocks. This transformation is <u>identical to the result previously derived for Displacement Current</u> (rotational magnetic field), where $H2'$ and $H3'$ are apparent field values deduced by observers in B arising from B's real space-time changes. So, observers in B measure the fields $H2'$ and $H3'$ to be non-zero, *even when the current is at rest relative to B*. <u>This resolves the Paradox of Magnetism</u>.

Because A is at rest relative to the external universe, and magnetism arises from a current's (absolute) motion relative to the external universe, $u1 = u1_p$ (the current's motion relative to A is also its absolute motion). The motion of the current relative to B plays no part in generating magnetism for B's observers.

Therefore, the motion $u1'$ of the current relative to B can be factored out by B's observers to give the absolute current velocity $u1$. If:

$$u1' = (u1 - v_p) / [1 - u1.v_p/c^2]$$
Then: $u1 = (u1' + v') / [1 + u1'.v'/c^2]$

Where $v' = v_p$, v' being the (relative) velocity of A relative to B. Although these two velocities are numerically equal, v' does not generate relativistic changes because it is relative motion (while v_p is absolute).

Relativistic Mass and Machian Electrodynamics

For system A, the coil and magnet interact by moving relative to each other. The same is determined by system B, but which also (like the coil and magnet) moves relative to both the external universe and A. However, the fields deduced by B are apparent (due to its space-time changes), and the true field values are those relative to A (e.g. relative to the external universe). So it can be seen that although induction arises by relative motion between the coil and magnet, the external universe still plays a role in the process.

Now, let's try to extend this reasoning further. So far, the coil and magnet are assumed to be (approximately) at absolute rest in A. System B has absolute motion, and undergoes asymmetric change relative to A. But what happens if the coil and magnetic have significant absolute motion too? The analysis so far relies on space-time changes in B. If the coil and magnet have absolute motion, the effects of relativistic mass increase need to be considered. Let the coil and magnet be a third system C with absolute velocity w_p relative to A. As determined by observers in A, we can write for any electromagnetic force F in C:

$$a(w_p) = \gamma(w_p).F \: / \: m_0$$

Where $m = m_0 \: / \: \sqrt{(1 - w_p^2/c^2)}$ is the relativistic mass increase of a body in C due to its absolute motion.

This equation can be re-scaled; assuming instead that force varies while mass is constant:

$$a(w_p) = \gamma(w_p).F_0 \: / \: m = F(w_p) \: / \: m$$

Where $F(w_p)$ is the apparent force due to relativistic mass increase. Assume that the apparent electromagnetic forces are proportional to their respective apparent electromagnetic fields, $F \propto E$ or $F \propto H$. Thus, relative to observers in A, the apparent electromagnetic fields in C are:

$$(E1, E2, E3) = \gamma(w_p).(E1_0, E2_0, E3_0)$$
$$(H1, H2, H3) = \gamma(w_p).(H1_0, H2_0, H3_0)$$

Relativistic simultaneity does not affect system A because it is at absolute rest. These equations can then be used in conjunction with the Machian Electrodynamic Equations to calculate how these fields vary relative to system B whose space-time has changed due to absolute motion relative to distant matter.

If the coil and magnet are relativistic, but have a different absolute velocity to that of B, then they constitute a third system whose dilation effects on induced currents need to be separately determined. However, for most situations in the Theory of Machian Electrodynamics, the motion of the coil and magnet will approximate to that of the earth, which itself approximates to a system at absolute rest.

Mach's Principle and Charge Density

In Maxwell's Equations, the charge density in a vacuum is given by:

$$\text{div } E = \rho$$

Where div is the divergence of the operator, E is the electric field strength, and ρ is 4π times the charge density (charge per unit volume). For a cylinder of charge Q, length L_0 and cross-sectional area A, at absolute rest, we have:

$$\text{div } E_0 = 4\pi \, Q \, / \, AL_0$$

If this cylinder is then given absolute motion along its axis, its length relative to stationary observers is $L = \gamma L_0$. Thus:

$$\text{div } E = 4\pi \, Q \, / \, A\gamma L_0 = \text{div } E_0 \, / \, \gamma$$

Relative to observers moving with the rod, it is at relative rest and there is no relative change in the rod's length, thus:

$$\text{div } E_0' = \text{div } E_0$$

However, relative to the moving observers, a stationary charged rod is lengthened, so we have instead:

$$\text{div } E' = \gamma . \text{div } E_0'$$

Conclusion

The application of Mach's Principle to Maxwell's Electrodynamics has been demonstrated. If my reasoning is disputed, and the covariance of inertial systems is asserted as true, then unambiguous evidence for symmetric (reciprocal) time dilation during inertial motion needs to be provided. Secondly, asymmetries must occur anyway for accelerated systems, so the asymmetry issue cannot be avoided. So, the equations I have developed must still apply when a system is accelerated.

But scientists and authors merely assume the asymmetry. Such as, for a round trip, taking the covariant equations for two systems, and ignoring the set for the accelerated system. In addition, the remaining equations are rarely presented in an asymmetric way. For example, if observers in inertial system A measure $t_B = \gamma . t_A$, then (by algebra) the corresponding equation for observers in accelerated system B should be $t_A = t_B / \gamma$. As for what happens during accelerated motion, cryptic references to General Relativity are usually given. What is more, how acceleration might apply to non-covariance in other phenomena, such as Maxwell's Equations, is never discussed.

However, I can derive the asymmetry from first principles, using properly asymmetric equations which unambiguously obey the rules or logic and algebra. And I have applied this to the equations of space-time (Lorentz–Einstein) and electromagnetism (Maxwell).

Glyn Phillips

PI

CONCLUSION

Once you have eliminated the impossible, whatever remains, no matter how improbable, must be the truth. (S. Holmes)

There is not a shred of experimental evidence for reciprocal time dilation during inertial motion as predicted by Special Relativity. This issue has never been previously raised by scientists. So far, experiments to test for time dilation have delivered only asymmetric results. Therefore Einstein's hypothesis of symmetry for inertial frames (the Relativity Principle) has not been specifically tested. So experiments need to be performed to test the hypothesis. Like I have said, if any reason is given that the hypothesis cannot be tested, then according to Popper's Principle, Special Relativity is not a proper scientific theory (pseudoscience) because it cannot be falsified. Possible excuses that might be used are "accelerations cannot be completely removed", or "symmetric results cannot be distinguished from asymmetric ones". According to the Theory of Machian Relativity, the results of such an experiment will still be asymmetric, as both inertial motion and acceleration always have asymmetric time dilation. This is simpler than conventional science, which has to alternate between the asymmetry of General Relativity and the symmetry of Special Relativity every time the motion changes from inertial to accelerated, or vice versa. According to the scientific principle of Occam's Razor, the simpler of two competing ideas is the more preferable. On this basis, Machian Relativity is more likely than Special Relativity to be correct.

In fact, the existing experimental evidence and theoretical arguments in favour of asymmetric dilation during inertial motion are so overwhelming and compelling that there is the case the experiments don't actually need to be done. And the belief by

167

scientists and authors that the Hafele-Keating Experiment proves Special Relativity alone, when in fact it also proves alternative theories (e.g. Aether Relativity and Machian Relativity), is a form of *confirmation bias*. Most scientists and authors will undoubtedly refuse to consider that Einstein might have been wrong in his formulation of Special Relativity. It is easier to defend the status quo and crush new ideas with cynicism. For example:

a) Finding a trivial error and then claiming it invalidates everything else in the book.

b) Misinterpreting some small point and then trying to "clarify" it with space-time diagrams and complex calculations based on General Relativity which few people are able to understand.

c) Trivialising the author by claiming he cannot know more than the "experts" because he is an amateur who doesn't work in a university, or that he is a crackpot or crank for daring to contradict theories that have "stood the test of time".

d) Refusing to publish any of the author's ideas because they have not been peer reviewed.

e) Rubbishing this book because it is self-published (without bothering to actually read it).

f) Claiming there is not one worthwhile idea in this book of over 200 pages, even though it has been written by someone who studied the subject at university (but then using some of its ideas in their own arguments).

g) Claiming Special Relativity must be correct because of time dilation experiments and GPS satellites, even though none of these things require reciprocal dilation.

It must be remembered that the issue is not about Machian Relativity but Special Relativity. Einstein was wrong about other things, so he could have been wrong about Special Relativity too. Like Einstein, the scientific community is claiming that relatively moving inertial clocks work slower than each other. Now, that is an extraordinary claim, and extraordinary claims require extraordinary levels of proof. So, professors in university physics departments need to be asked if there is any unambiguous experimental evidence for reciprocal time dilation during inertial motion. If it is claimed that the evidence exists, then the world needs to hear about it. So, all available physics books will have to be republished with this data. Also, this book is not just about the relativity of space-time, but other aspects too, such as Machian Inertia and Electrodynamics. There are also all my other proofs and theorems. So, if scientists and authors wish to dismiss this book as the ramblings of an amateur crackpot, then *they need to refute all of its ideas, and not just one part of it.*

Scientists and authors need to decide which of their "explanations" of the Twins Paradox (i.e. time dilation fudge factors) is the definitive one. They cannot all be correct. They should then provide experimental evidence (from atomic clocks performing round trips) to provide confirmation of their particular choice. But such Twins Paradox "explanations" are more like an *argumentum ad nauseam* – because there are so many of them, people are bamboozled into thinking Einstein's theory must be correct. However, Machian Relativity does not need supplementary explanations to save it. Unlike Special Relativity, Machian Relativity fully incorporates Mach's Principle, because it includes the relative effects of mass in the universe. This means relativistic effects are absolute rather than purely relative, because mass effects act in combination, producing a universal average. But otherwise the theory is as close as possible to Einstein's. It incorporates his mathematical methods for calculating time dilation, the speed of light is constant and $E = mc^2$. And because the apparent effects of time delays and relativistic simultaneity are properly factored out, the true physical effects can be determined. Even if scientists claim that inertial motion cannot have asymmetries, such asymmetries must still exist during acceleration, and this book describes how they are mathematically derived.

In understanding Special Relativity, it is said you cannot rely on common sense. But Machian Relativity has common sense, because it obeys the rules of logic, algebra and physical change. So science now has a choice between two theories; one with common sense, and the other without it.

APPENDIX I: SUMMARY OF THE TWO THEORIES

Special Relativity[10]

System A (inertial)	System B (inertial)

$$T_B = T_A \sqrt{(1 - v^2/c^2)}$$

$$L_B = L_A \sqrt{(1 - v^2/c^2)}$$

$$t_B = -vx_A / c^2\sqrt{(1 - v^2/c^2)}$$
$$t_A = 0$$

$$T_A = T_B \sqrt{(1 - v^2/c^2)}$$

$$L_A = L_B \sqrt{(1 - v^2/c^2)}$$

$$t_B = 0$$
$$t_A = vx_B / c^2\sqrt{(1 - v^2/c^2)}$$

System A (inertial)	System B (non-inertial)

$$T_B = T_A \sqrt{(1 - v^2/c^2)}$$

$$L_B = L_A \sqrt{(1 - v^2/c^2)}$$

$$t_B = -vx_A / c^2\sqrt{(1 - v^2/c^2)}$$
$$t_A = 0$$

$$T_A = T_B / \sqrt{(1 - v^2/c^2)}$$

$$L_A = L_B / \sqrt{(1 - v^2/c^2)}$$

$$t_B = -vx_A / c^2\sqrt{(1 - v^2/c^2)}$$
$$t_A = 0$$

---

[10] Note: In Special Relativity, the asymmetry of the second set of equations (when B is non-inertial) is assumed by arbitrarily inverting the relativistic coefficient $\sqrt{(1 - v^2/c^2)}$ to $1 / \sqrt{(1 - v^2/c^2)}$ – the inversion is not actually derived mathematically from first principles per-se. Conventional science only describes asymmetric length contraction (L) and time dilation (T) during acceleration, but gives no corresponding description of what happens to relativistic simultaneity (t_A, t_B). I have assumed that for this theory, relativistic simultaneity for the non-inertial frame persists during acceleration, and is asymmetric relative to the inertial frame, so that it is objectively true for observers in either frame.

Machian Relativity[11]

System A (absolute rest)	System B (absolute inertial motion)

$T_B = T_A \sqrt{(1 - v^2/c^2)}$

$T_A = T_B \sqrt{(1 - v^2/c^2)}$
(apparent)

$T_A = T_B / \sqrt{(1 - v^2/c^2)}$
(true)

$L_B = L_A \sqrt{(1 - v^2/c^2)}$

$L_A = L_B \sqrt{(1 - v^2/c^2)}$
(apparent)

$L_A = L_B / \sqrt{(1 - v^2/c^2)}$
(true)

$t_B = -vx_A / c^2\sqrt{(1 - v^2/c^2)}$
$t_A = 0$

$t_B = 0$
$t_A = vx_B / c^2\sqrt{(1 - v^2/c^2)}$
(apparent)

$t_B = -vx_A / c^2\sqrt{(1 - v^2/c^2)}$
$t_A = 0$
(true)

---

[11] Note: Machian Relativity obeys algebra and is paradox-free because the changes are asymmetric, while the covariance is apparent. It applies to both inertial motion and acceleration by applied forces along any linear, curved or polygonal line. Here "v" is absolute velocity, not relative velocity. This arises due to the presence of matter in the External Universe (the "Fixed Stars"), motion relative to which causes physical changes. In Machian Relativity, relativistic simultaneity is not an inherent property of time, so will not occur in an accelerated system. In this case see Appendix VI for the corresponding transformations (based only on asymmetric L and T). When acceleration stops and a system achieves absolute inertial motion, its observers can recalibrate its clock readings according to the above equations (t_A, t_B) for relativistic simultaneity.

APPENDIX II: THE GALILEAN TRANSFORMATIONS
PRESERVE THE FORM OF MAXWELL'S EQUATIONS

Maxwell used his electromagnetic equations to predict the existence of electromagnetic waves that travelled at speed c relative to the Aether. Relative to a coordinate system (x_E, t_E) at rest in the Aether, this is:

$$\partial^2 E/\partial x_E^2 = 1/c^2. \; \partial^2 E/\partial t_E^2$$

Where; E is the electric field strength. Maxwell formulated his theory before the advent of relativity, so he assumed the Galilean Transforms were still valid. The form of this equation for an observer moving relative to the Aether, assuming these transformations, will be explained later.

In Lorentz's Aether Relativity, for a system A at rest in the Aether, and a system B moving at absolute speed v relative to the Aether, we have:

$$\partial^2 E/\partial x_A^2 = 1/c^2. \; \partial^2 E/\partial t_A^2$$

$$\partial^2 E/\partial x_B^2 = 1/c^2. \; \partial^2 E/\partial t_B^2$$

This shows that the invariance of the speed of light applies to the wave equation (as well as the space-time equations), and that the form of the equation is preserved (covariance). The same reasoning applies to Special Relativity and Machian Relativity. In Special Relativity, the speed v is the relative speed (v_R) between A and B, and the covariance of the wave equation is also assumed to apply to the space-time of both systems ("there is no absolute motion"). In Lorentzian and Machian Relativity, the speed v refers only to the absolute motion of system B, relative to the Aether (v_E) or relative to the Plenum/external universe (v_P) respectively. In these two theories, the covariance of the wave equation only means the speed of light is invariant for both systems, and this symmetry does not extend to the space-time of the systems [This gives a far simpler explanation for the Twins Paradox than Einstein's method].

Corollary: Electromagnetic Waves in Galilean Relativity

The transformation of Maxwell's wave equation between different systems, using the Galilean Transform equations will now be considered. In system A let there be an electromagnetic wave of the form:

$$\partial^2 E/\partial x_A^2 = 1/c^2 \cdot \partial^2 E/\partial t_A^2$$

Let system B move at inertial speed v relative to A. If instead of the Lorentz Equations, the Galilean Transforms are used instead to calculate how this wave appears in system B, the following equations are generated:

$$\partial^2 E/\partial x_A^2 = \partial^2 E/\partial x_B^2 + \{2/v\}\partial^2 E/\partial x_B \partial t_B + \{1/v^2\}\partial^2 E/\partial t_B^2$$
$$\partial^2 E/\partial t_A^2 = \partial^2 E/\partial dt_B^2 - \{2v/c^2\}\partial^2 E/\partial x_B \partial t_B + \{v^2\}\partial^2 E/\partial x_B^2$$

Because of the additional terms, it is said that the Galilean Transformation "does not preserve the form of the wave equation", and that there is no experimental evidence for the additional terms. While the latter is true, I say that the former statement is specious. I will now show that these additional terms arise because the equations wrongly assume the speed of light to remain invariant in Galilean Relativity.

To resolve this issue, consider a transverse wave in water or on a string. Speeds are non-relativistic, so the Galilean Transformations can be used. Relative to a stationary observer, if the wave has speed w, the wave equation is:

$$\partial^2 \Psi/\partial x_A^2 = 1/w^2 \cdot \partial^2 \Psi/\partial t_A^2$$

Where; Ψ is the wave amplitude. Let a second observer B move at speed v relative to A. For this second observer, the wave equation is now:

$$\partial^2 \Psi/\partial x_B^2 = 1/(w-v)^2 \cdot \partial^2 \Psi/\partial t_B^2$$

So, if B moves at the same speed as the wave (e.g. a surfer), the wave appears static. Therefore, the Galilean Transformation preserves the wave equation form for transverse waves in water or on a string. Thus, it would be expected that the same reasoning would apply to an electromagnetic wave using the Galilean transforms. According to Galilean Relativity, because space-time is invariant, the speed of light is not constant. [We know this is false, but things must be done in accordance with Galilean Relativity for a consistent analysis]. Using the Galilean Transforms, if the speed of light relative to A is c, the corresponding speed (of the same light) relative to B is:

$$c_B = c - v$$

When observers in A measure $\partial E/\partial x_A$, this is done simultaneously between two points in A. Because of the simultaneity, $\partial E/\partial t_A = 0$. And because simultaneity is preserved between systems in Galilean Relativity (unlike in the other theories that use the Lorentz equations), we also have $\partial E/\partial t_B = 0$ in system B. Thus, terms with $\partial/\partial t_B$ cancel out, giving:

$$\partial^2 E/\partial x_A{}^2 = \partial^2 E/\partial x_B{}^2$$

Similarly, when observers in A measure $\partial E/\partial t_A$, this must be done at a point in A, and the same applies to observers in B. Therefore, $\partial/\partial x_A = \partial/\partial x_B = 0$. This gives:

$$\partial^2 E/\partial t_A{}^2 = \partial^2 E/\partial t_B{}^2$$

Therefore, the Galilean transforms give the following equations for a transverse electromagnetic wave:

$$\partial^2 E/\partial x_A{}^2 = \qquad 1/c^2 . \, \partial^2 E/\partial t_A{}^2$$
$$\partial^2 E/\partial x_B{}^2 = 1/(c-v)^2 . \, \partial^2 E/\partial t_B{}^2$$

The first equation can be considered as a special case of the second, where $v = 0$. The same equations apply in Maxwell's Aether Theory if A is at rest in the Aether. However, in Galilean Relativity (no Aether), for a source at rest in B we instead have $1/c^2$ relative to B and $1/(c+v)^2$ relative to A. Hence, the claim that the Galilean transforms do not preserve the form of the electromagnetic wave equation is false, because the additional terms cancel. However, because of the Michelson-Morley Experiment, it ultimately follows that the Galilean Transforms are invalid, and the Lorentz Equations[12] must be used instead (but not in the symmetric way as Einstein believed). Also, relativistic mass increase means bodies cannot reach the speed of light. The Lorentz Transformation is fundamental, and the Galilean Transformation is only as a non-relativistic approximation.

---

[12] Further mention needs to be made of the different emission processes in each theory. In *Maxwell's theory*, the motion of light is always absolute relative to the ether, regardless of the source motion. Thus for a system moving relative to ether, its observers determine light to have speed c-v, even for sources moving with the system. For *Newtonian emission*, light moves at c relative to its source (there is no ether), as long as the sources continue to move inertially. Therefore, the speed of light from any relatively moving light source is c-v. For *Ritzian emission*, light moves at c relative to its source, even when the source is accelerated. This obviously requires a causal space around the source (e.g. a Mach Field). But here, the influence of other Mach Fields on the light rays is not included. *Lorentz emission* is similar to Maxwell's, but the physical change of a system moving inertially relative to the Aether means the speed of light for the system (from any source) is constant. *Machian emission* is similar to Lorentz's, but the role of the Aether is replaced by the Mach fields of external matter. In *Special Relativity*, neither the Aether nor Mach's external matter has a role. Instead, different inertial systems change paradoxically relative to each other, and the motion of wavefronts from different sources are paradoxically centred on any source that is arbitrarily considered to be "at rest". So, the nature of the emission theory is not adequately resolved (e.g. whether each source acts as a local "ether" for the wavefronts from other sources or whether each source is a Newtonian type emitter). In addition, Special Relativity does not give a clear answer as to what happens when a source is accelerated. In *Stokes's Aether Drag Theory*, light has absolute motion relative to the Aether, except near massive bodies, when its absolute motion is instead relative to such bodies.

APPENDIX III: PARADOX-FREE TIME DILATION IN MACHIAN RELATIVITY

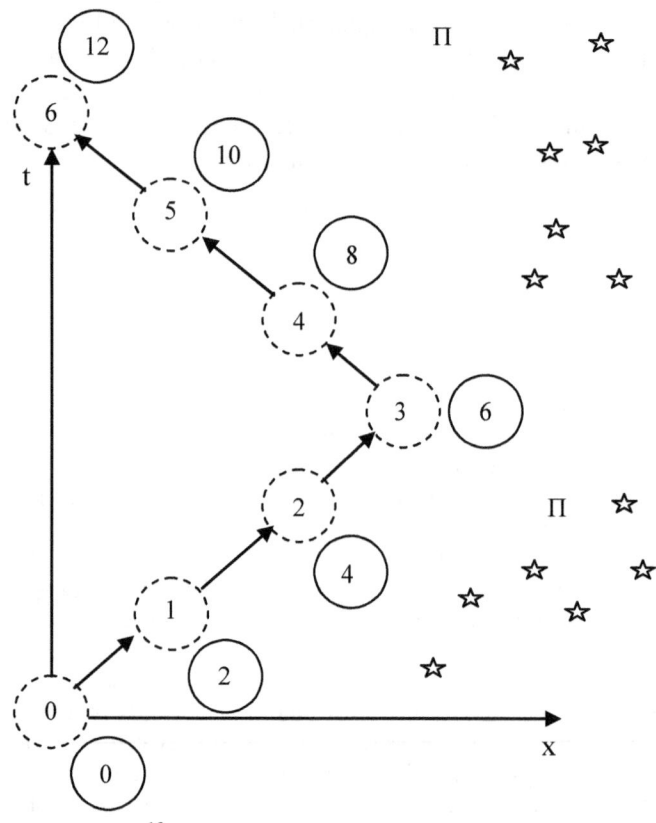

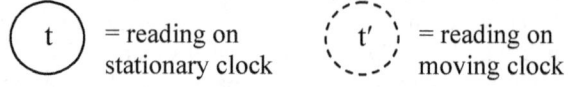

Fig.41: Readings[13] on Moving Clock and Stationary Clocks

$\left(\; t \;\right)$ = reading on stationary clock

$\left(\; t' \;\right)$ = reading on moving clock

---

[13] Note: These clock readings are objectively true for observers with the moving clock and the stationary clocks. The time dilation of the moving clock is caused by its absolute motion through the Mach fields of the external universe (Π). In this example, the absolute velocity of the moving clock is such that $\gamma = 0.5$.

APPENDIX IV: WAVEFRONTS FROM SOURCES WITH ABSOLUTE MOTION OR AT ABSOLUTE REST

$\oplus$ = stationary source $\otimes$ = moving source

$\Psi_{1,2,3}$ = wavefronts from $\oplus$ $\Psi'_{1,2,3}$ = wavefronts from $\otimes$

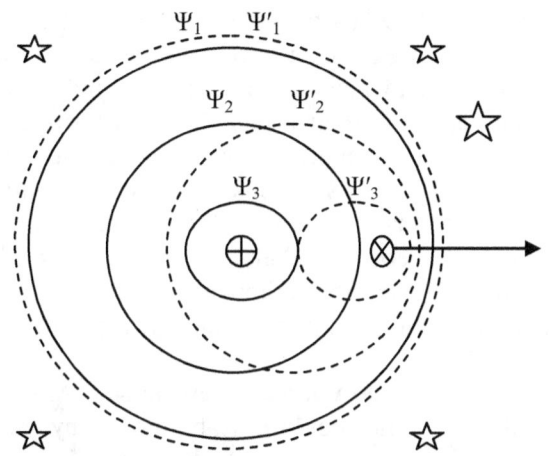

Fig.42: Wavefronts Relative to Stationary System

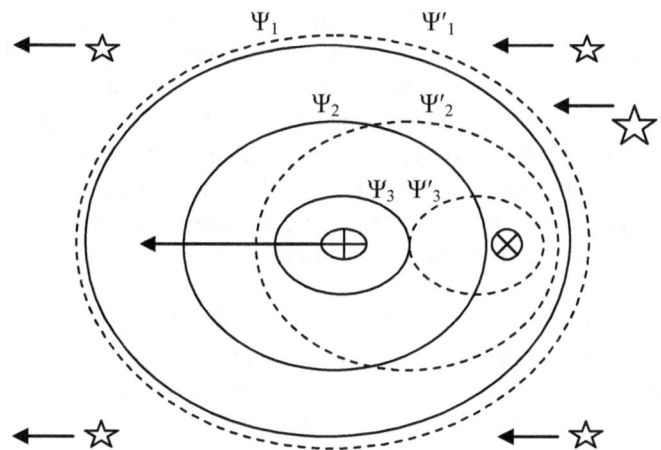

Fig.43: Wavefronts Relative to Moving System

APPENDIX V: GENERAL RELATIVITY DOES NOT EXPLAIN MACH'S PRINCIPLE

Descartes, Newton and Berkeley

Isaac Newton, in his book Principia, described what he called Absolute Space, which he said was "immovable". This was in response to Rene Descartes, who explained planetary motions in terms of his Vortex Theory. Space was a movable substance, allowing the sun to generate a vortex around itself. This vortex carried the planets in their orbits. Therefore, space was not a true void (even for a vacuum), in that it was filled with a kind of substance. In Newton's mechanics, the planetary orbits were explained in terms of Universal Gravity, based on forces of attraction. This theory was better than Descartes' Vortex Theory, because it predicted the elliptical orbits of comets, and also explained the motion of the moon and gravity on earth.

Because there were no vortices, Newton said Absolute Space was immovable. So bodies could only be moved by forces, according to Newton's First Law (inertia); "a body remains at rest, or moves in a straight line at constant speed, unless acted on by a resultant force". This explained non-inertial effects during rotation or linear acceleration – if you are accelerating, bodies in your vicinity (like the rest of the universe) do not move with you. Instead, they have relative acceleration due to their constant motion in Absolute Space (while you accelerate in Absolute Space, due to interaction with other matter according to Newton's Third Law of action and reaction). Relative to you, other bodies experience "fictitious forces" because they are not mutually interacting. Newton also said "space and time are absolute", meaning measuring rods and clocks were not affected by motion relative to Absolute Space.

To illustrate his idea of Absolute Space, Newton described his *Bucket experiment*. A bucket is attached to a rope, which is then twisted. The bucket is filled with water and is then released, so that it begins spinning due to the untwisting rope. Initially the water remains at rest and is flat, but after a while friction with the spinning bucket causes the water to rotate. After a while, the water

Glyn Phillips

rotates with the bucket and has a concave surface. Newton argued that the water's concave surface was due to its rotation relative to Absolute Space rather than the bucket, because if the bucket was suddenly stopped, the water would continue rotating for a while with the curved surface (Newton's intention is to show that the bucket only imparts motion to the water but does not cause the concave surface).

Bishop George Berkeley objected to the idea of Absolute Space, because apart from its effects it cannot be directly seen, so he developed his Fixed Stars idea. Motion is only relative, because we can only define a body's motion relative to other perceivable bodies. Thus, a body caused to rotate in the absence of the Fixed Stars would experience no centrifugal effect, but with the Fixed Stars present, its parts would tend to move towards such stars, causing centrifugal effects. Berkeley's idea of acceleration is not truly relative, as the Fixed Stars remain immovable. What he means by "relative" is motion relative to the perceivable Fixed Stars, rather than Absolute Space. He doesn't explain how stars exert influence across space – they must cause a field like Absolute Space to do this. Berkeley's suggestion of what might happen in an empty universe is untestable and probably wrong – applying forces to make a body rotate would require mutual interaction with some other body, thereby providing an alternative system other than the stars for acceleration effects. Newton's original aim was to refute the idea that the bucket by itself affects the water, rather than speculating on the cause of Absolute Space. As he said in his Principia, "Hypotheses non fingo" ("I do not devise hypotheses").

Ernst Mach

In Newtonian Relativity, absolute velocity has no effect during inertial motion. It is only noticeable during non-inertial motion (such as rotation). For inertial motion, we can only know a body's motion relative to other inertial bodies, rather than Absolute Space per-se. So Mach sought to extend this to non-inertial motion, so that non-inertial effects arose purely as a result of acceleration relative to inertial bodies.

179

Although it had been discovered that the "Fixed Stars" were not actually fixed, Mach reasoned that such stars, being sufficiently remote from each other, were inertial bodies relative to which non-inertial motions could be ascertained.

Like Berkeley, Newton's Bucket result was due to the bucket being caused to rotate relative to the Fixed Stars. But for Mach, this relative effect was a *general rule*, so that if the Fixed Stars were instead caused to rotate around the bucket, the water would experience exactly the same centrifugal effect. In Newton's scheme, rotating the stars around the bucket would have no effect on the water, because it is stationary relative to Absolute Space. According to Mach, because acceleration is purely relative, we cannot objectively know whether the bucket or the stars are really rotating.

However, Mach took his ideas too far. *Firstly*, it cannot be tested, because we cannot rotate stars around the bucket, so Mach's suggestion is more like Metaphysics. *Secondly*, if it could be done, it would require interaction with other matter, generating an equal and opposite angular momentum to that of the stars, thereby nullifying any centrifugal effect on the bucket. *Thirdly*, Mach suggested the water might not be concave if the walls of Newton's Bucket were "made leagues thick". That is, massive bodies might produce localised inertial dragging, overcoming the influence of the Fixed Stars. But this is implausible, as the dragging variation with distance requires explanation. Also, no other phenomena show dragging effects near the earth - stellar aberration and the time dilation of clocks on the rotating earth both refute localised dragging (and Einstein's idea that earth observers can consider themselves "at rest"). Neither can this be used to criticise Newton, as his original intention was to refute the erroneous notion that a normal bucket *by itself* might somehow be causing the centrifugal effects. *Fourthly*, while inertial systems are equivalent to each other, non-inertial systems are not equivalent to inertial ones, no matter how such non-inertial effects are explained. This is what makes non-inertial systems Absolute per-se, even if there is no Absolute Space as such.

Albert Einstein

Einstein continued Mach's efforts to explain acceleration as being relative and tried to incorporate Mach's Principle into his General Relativity. He began with Mach's assertion that non-inertial effects were somehow due to the presence of matter in the Universe. However, he then *digressed from Mach* and introduced his Equivalence Principle, where non-inertial effects are described in terms of an accelerated system being "at rest" in an induced gravitational field. Supposedly this abolished the distinction between absolute rest and motion during acceleration, making *acceleration per-se* a purely relative effect. This (supposedly) unified accelerated motion with Special Relativity, which had abolished the Aether (making inertial motion a purely relative effect). However, unlike in Special Relativity, there is no complete reciprocity between inertial and non-inertial coordinate frames – non-inertial frames still experience non-inertial effects while inertial frames don't, and neither is there reciprocal time dilation between them (so this method appears to be a dead end).

There are other problems. *Firstly*, the "Fixed Stars" are not causing non-inertial effects – the cause is the induced gravitational field, in which the "Fixed Stars" passively fall. So it doesn't incorporate Mach's Principle, and Einstein has taken us on a wild goose chase. *Secondly*, when applied to rotating systems, the Equivalence Principle predicts an outward (centrifugal) gravitational field. However, this contradicts the rotation of matter external to the system, which requires an inward (centripetal) force. *Thirdly*, the induced fields violate Newton's Third Law - during linear acceleration, matter "falling" in the induced gravitational field is not mutually interacting with other matter.

Einstein later claimed to have found solutions in his General Relativity that were Machian in nature. But these are localised dragging effects near masses, not the external universe as a whole. His theory still requires induced gravitational fields to explain non-inertial effects, not the Fixed Stars. So General Relativity does not include Mach's Principle.

Machian Relativity

Non-inertial effects are explained by Machian Relativity – bodies move inertially relative to the *Average Momentum of the Universe*, unless acted on by a force. All bodies have Mach Fields, which fill all space forming a Plenum. This explains how the "Fixed Stars" act across space. The average momentum of the Universe is the frame in which the total momentum of the Universe is zero, and defines a state of absolute rest. This is because the total momentum of mutually interacting bodies is conserved according to Newton's Third Law, and thus the total momentum of the Universe (being a closed system) is conserved. If, as Mach imagined, the Fixed Stars were rotated around Newton's Bucket, this would require interaction with other matter, generating an equal and opposite angular momentum. Hence there will be no centrifugal effect on the bucket. Hence, Mach's belief that the presence of the Fixed Stars leads to relative acceleration is false. On the contrary, the presence of the Fixed Stars means acceleration is Absolute.

And Machian Relativity shows how absolute inertial velocity has an *objective meaning*. The same Mach Fields that determine non-inertial effects also determine the absolute motion of light and absolute space-time changes in all frames, including those moving inertially. Hence Time Dilation and Length Contraction are asymmetric and relative to the average momentum of the Universe. This gives the simplest explanation for the Twins Paradox – the travelling twin ages asymmetrically during every part of his journey. Similarly, the Michelson-Morley null result is due to a *physical contraction* of the apparatus due to the earth's absolute motion, which counters the "Machian wind" effect. In the Hafele-Keating Experiment, clocks on earth are physically time dilated due to the earth's rotation, proving that the earth cannot be "at rest" in a relative sense. An uncorrected clock orbiting the earth in gravitational free-fall must also be asymmetrically dilated, even though it is locally inertial. Hence there can be no epistemological objection to a clock being asymmetrically dilated when moving inertially relative to the earth through space, which is not a void, but a Plenum of the Gravitational and Mach Fields of all matter.

APPENDIX VI

HOW ASYMMETRIES GENERATE APPARENT
COVARIANCE AND HOW IT IS FACTORED OUT

Consider two relatively moving inertial frames. Let frame S be at
absolute rest, which is determined by the *average momentum of all
matter in the universe, according to Mach's Principle*. Let frame
S' have inertial velocity v relative to the first, which due to Mach's
Principle, is a state of absolute motion. Also, let the Universal
Reference Frame for all light (according to De Sitter's Binary Star
Principle) correspond to this state of absolute rest:

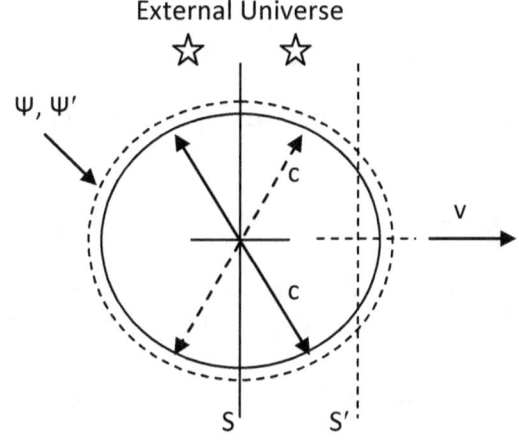

Fig.44

Because S is at absolute rest, light spheres from both its own
source (Ψ) and that in S' (Ψ') remain centred on its coordinate
origin. So the space-time of S' needs to change to maintain c-
invariance for these light spheres. That is, clocks and rulers in S'
undergo physical changes due to its absolute motion, and this
keeps the speed of light invariant for its observers.

Both frames have their relative velocities along the x-axes, which
are aligned in the usual way. Observers in S make space-time
measurements (x, t) and observers in S' make measurements (x',

t'). These measurements are *objectively true* for observers in either system, and not just "how one system looks relative to observers in the other" – they look that way because that is how they actually are.

External Universe

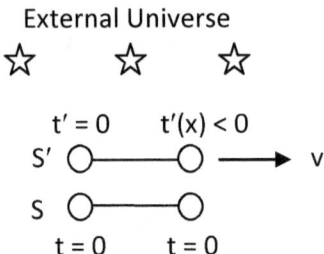

Fig.45: Absolute Motion of S' relative to S and "Fixed Stars"

The measurements made in S' are calculated from those in S, using the Lorentz Transforms:

$$x' = (x - vt) / \gamma \qquad t' = (t - vx/c^2) / \gamma \qquad \{1a\}$$

These predict that clock *rates* and ruler *lengths* in S' are time dilated and length contracted due to motion relative to S given by:

$$T' = T \sqrt{(1 - v^2/c^2)} \qquad L' = L \sqrt{(1 - v^2/c^2)} \qquad \{1b\}$$

In addition, the *reading* of the clock in S' adjacent to x in S is shifted by relativistic simultaneity according to:

$$t'(x) = -vx / c^2\sqrt{(1 - v^2/c^2)} \qquad \{1c\}$$

Now, *contrary to Special Relativity, let us imagine that these changes in S' are physical in nature, due to its absolute motion.* Therefore, relative to observers in S', clock rates and ruler lengths in S are instead given by:

$$T = T' / \sqrt{(1 - v^2/c^2)} \qquad L = L' / \sqrt{(1 - v^2/c^2)} \qquad \{2b\}$$

Glyn Phillips

That is, these new equations {2b} <u>invert</u> the relativistic coefficient in {1b} and <u>are not covariant</u> with those of {2b}, but are instead a straightforward algebraic rearrangement of them. That is, both sets of equations {1b and 2b} now obey the rules of algebra, logic and physical change. If the equations {1b} are called "time dilation" and "length contraction", we might want to give {2b} names too, for example "*inverse dilation*" and "*inverse contraction*". In S' all clocks and rulers are equally affected, so observers determine no relative change within their own frame. But, this does not imply that nothing has happened to the space-time of S' or that S must change reciprocally to S'.

The change in S' is <u>absolute</u>. The corresponding inverse changes in S (relative to observers in S') are <u>relative</u> only. S does not really change, and the true cause lies within S' itself. *Time dilation* maintains c-invariance along the y'-axis of the moving frame (e.g. as in light clocks). *Length contraction* maintains c invariance for light moving along the x'-axis from the origin and reflected back (e.g. the Michelson-Morley experiment). *Relativistic simultaneity* maintains c-invariance for light on the x'-axis which is not reflected. The covariant equations for observers in S' are:

$$T = \underline{T' \sqrt{(1 - v^2/c^2)}} \qquad L = \underline{L' \sqrt{(1 - v^2/c^2)}} \qquad \{3b\}$$

$$t(x') = \underline{vx' / c^2\sqrt{(1 - v^2/c^2)}} \qquad \{3c\}$$

It will now be shown that this covariance is an <u>apparent effect</u>, arising from absolute and objective simultaneity changes in S'.

<u>Apparent Covariance - Relativistic Simultaneity</u>

So, how can this asymmetric scenario generate covariance? The answer lies with relativistic simultaneity. Let all clocks in S be synchronised to read t = 0 when the origins of both systems coincide. Let clocks for S' have relativistic simultaneity according to the usual equation of Special Relativity:

$$t'(x) = \underline{-vx / c^2\sqrt{(1 - v^2/c^2)}}$$

185

So t' is a function of x; $t'(x)$. Because these clock readings are objectively true, t' is also a function of x', where $x' = x / \sqrt{(1 - v^2/c^2)}$.

Let these clock readings for S' be <u>objectively true for observers in either system</u>. The same objectivity applies to clock readings in S. In other words, the relativistic simultaneity for S' is absolute in nature. Therefore, when the origins of both systems coincide, the reading of the clock at the origin of S' is $t' = 0$ and the reading of another clock further along the x' axis (corresponding to x on the axis of S) is $t' = -vx / c^2\sqrt{(1 - v^2/c^2)}$; while the corresponding reading for an adjacent clock in S is $t = 0$.

Now examine what happens when the clock in S' advances by a (time dilated) duration $T' = vx / c^2\sqrt{(1 - v^2/c^2)}$, so that it reads $t' = 0$. The origins will no longer coincide, and S' will have travelled some distance relative to S. The corresponding readings for clocks in S will advance by a greater amount ("inverse dilation") given by:

$$t = T' / \sqrt{(1 - v^2/c^2)} \; = vx / c^2(1 - v^2/c^2)$$

So, a clock in S adjacent to $x' = x / \sqrt{(1 - v^2/c^2)}$ reads $vx / c^2(1 - v^2/c^2)$. We now need to express t as a function of x'. We know $x = x'\sqrt{(1 - v^2/c^2)}$, so substitution yields:

$$t(x') = v\, x'\, \sqrt{(1 - v^2/c^2)} / c^2 (1 - v^2/c^2)$$
$$= \underline{v\, x' / c^2\, \sqrt{(1 - v^2/c^2)}}$$

This equation is covariant with that for $t(x)$ and is the same as that predicted by Special Relativity {3c}, but the <u>physical interpretation is different</u>. In order for this scenario to happen, the system S' has had to travel a certain distance for the clock along its x'-axis to advance to read zero. So equalised synchronisation for two particular clocks in S' only happens at two instants, while in S it happens for all clocks and at all instants:

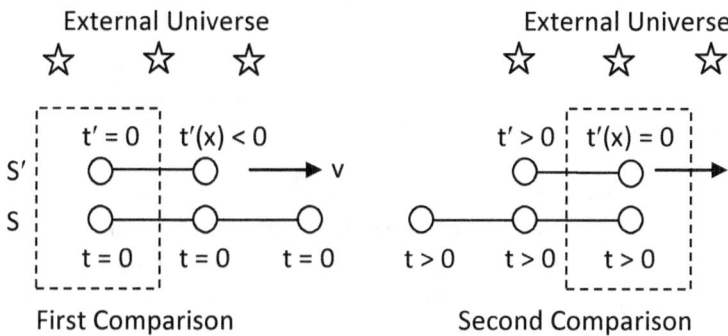

First Comparison   Second Comparison

Fig.46: Apparent Covariant Simultaneity

That is, the relativistic simultaneity in S' is *physical and absolute*, and the equivalent scenario in S (predicted by Special Relativity because there is no absolute motion), where $t = v\,x' / c^2\,\sqrt{(1 - v^2/c^2)}$ when the origins of both systems are coincident, <u>does not occur</u>. This is because of the influence of the External Universe (Mach's Principle). Thus only S' is changed because its motion is absolute (it is not at rest relative to the External Universe):

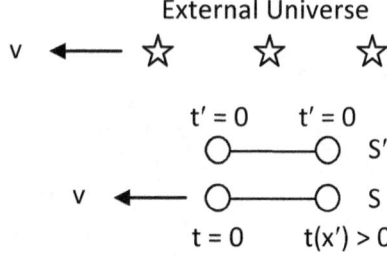

Fig.47: The Reciprocal Situation does not occur

In other words, the two systems <u>are not equivalent</u>. So when the origins coincide, observers in both systems objectively conclude clock readings are as follows:

$$t' = -vx / c^2\sqrt{(1 - v^2/c^2)} \qquad t = 0 \qquad \{2c\}$$

Apparent Covariance - Time Dilation

It has been shown how the absolute relativistic simultaneity of S′ creates an apparent relativistic simultaneity in S. The effects of this simultaneity on time dilation will now be investigated.

Consider the two frames as before, when their origins coincide. A clock on the x-axis of S reads $t = 0$. By relativistic simultaneity, the adjacent clock in S′ reads $-vx / c^2 \sqrt{(1 - v^2/c^2)}$. This reading is objectively true for observers in either system. Frame S′ then undergoes absolute motion relative to S, so that the origin of S′ reaches the clock at x. Clocks in S will advance to read $t2 = x / v$. Because of absolute time dilation, clocks in S′ advance by $\gamma . t2 = \gamma\, x / v$. So the reading of the clock the origin of S′ is now $\gamma\, x / v$. So the rate of the clock in S relative to clocks in S′ is:

$$(t2 - t1) = T = (x/v) - 0$$
$$(t2' - t1') = T' = \gamma\, x / v + vx / c^2 \sqrt{(1 - v^2/c^2)}$$

Thus:
$$T' / T = \gamma + v^2 / c^2 \sqrt{(1 - v^2/c^2)}$$
$$= \sqrt{(1 - v^2/c^2)} + v^2 / c^2 \sqrt{(1 - v^2/c^2)}$$
$$= 1 / \sqrt{(1 - v^2/c^2)}$$

Therefore:
$$T = \underline{T' \sqrt{(1 - v^2/c^2)}}$$

So it can be seen that this equation is covariant with the previous equation {1b} describing the time dilation of clocks in S′ due to their motion relative to S. But unlike in Special Relativity, this covariance is apparent, and arises from the relativistic simultaneity of clocks in S′ which is physical and absolute. The true rate of clocks in S relative to S′ is given by equation {2b}, where:

$$T = \underline{T' / \sqrt{(1 - v^2/c^2)}}$$

So it can be seen how the absolute relativistic simultaneity in S′ causes apparent covariance by inverting the relativistic coefficient $1 / \sqrt{(1 - v^2/c^2)}$.

Glyn Phillips

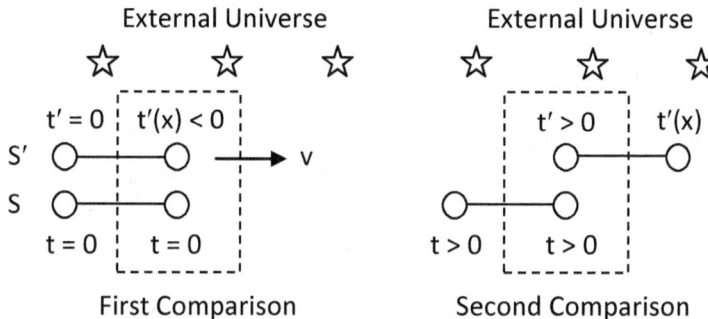

First Comparison   Second Comparison

Fig.48: Apparent Covariant Time Dilation

This explains the asymmetric outcomes of the *Twins Paradox* and *time dilation experiments* in the simplest possible way. The equations are inherently asymmetric, and this applies to both inertial and accelerated motion. This avoids the situation in conventional science of trying to get the symmetric Special Relativity to give an asymmetric result.

Apparent Covariance - Length Contraction

Now, when observers in both systems directly compare their rulers, the measurements made by observers in S′ are not simultaneous. This is because the clocks at either end of its ruler are non-synchronous due to relativistic simultaneity, which (according to the reasoning here) is absolute and objective for both sets of observers. However, what would happen if observers in S′ could somehow make a length measurement that kept these two clocks simultaneous?

Consider in the frame S′ a clock on the end of a ruler of length L whose end is at x1′. Because of length contraction of rulers in S′, this corresponds to x1 = γ x1′ in S. Therefore by relativistic simultaneity, the reading on this clock is:

$$t1' = -vx1 \,/\, c^2 \,\sqrt{(1 - v^2/c^2)}$$

189

Consider this as Event 1:
For S' we have: $\quad\quad\quad$ x1', t1'
For S we have: $\quad\quad\quad\quad$ x1, t1 $= \gamma$ x1', 0

Let this clock advance by vx1 $/ c^2 \sqrt{(1 - v^2/c^2)}$ so that it reads t2' = 0. Clocks in S advance (by inverse dilation) according to:

$$t2 = t2' / \gamma = vx1 / c^2 (1 - v^2/c^2)$$

So in this duration, S' travels a distance d relative to S by:

$$d = vt2 = v^2 x1 / c^2 (1 - v^2/c^2)$$

Therefore the new position of this clock relative to S is:

$$x2 = x1 + d = x1 + v^2 x1 / c^2 (1 - v^2/c^2)$$
$$= x1 / (1 - v^2/c^2) = x1 / \gamma^2$$

So for Event 2 we have:
In frame S' we have: $\quad\quad$ x2', t2' = x1', 0
In frame S we have: $\quad\quad$ x2, t2

Therefore when observers in S' make a length measurement x1' that preserves clock synchronisation, the corresponding measurement in S is x1 $/ \gamma^2$. We know that because of Lorentz contraction of frame S', x1 $= \gamma$ x1'. Therefore by substitution:

$$x2 = x1 / \gamma^2 = \gamma x1' / \gamma^2 = x1' / \gamma$$

So we need to express this result in terms of the lengths of objects in each frame. We know x1' in S' is the length of a ruler for observers in the frame, so that x1' = L'. Thus:

$$x2 = x1' / \gamma = L' / \gamma$$

So the corresponding measurement x2 is larger than the measurement x1' = L'. This implies rulers in S have contracted relative to those in S' according to:

$$\underline{L = \gamma\,L'}$$

This result is covariant with the length contraction of S' due to its motion relative to S (L' = γ L). However, like the result for time dilation, the covariance <u>is apparent</u>, and the relativistic simultaneity of S inverts the relativistic coefficient (1 / γ → γ). The clock at the end of the ruler in S' has to travel an extra distance for its clock to advance to read zero. This results in a greater reading in S, which implies rulers in S have contracted to give a greater distance reading, but this is false.

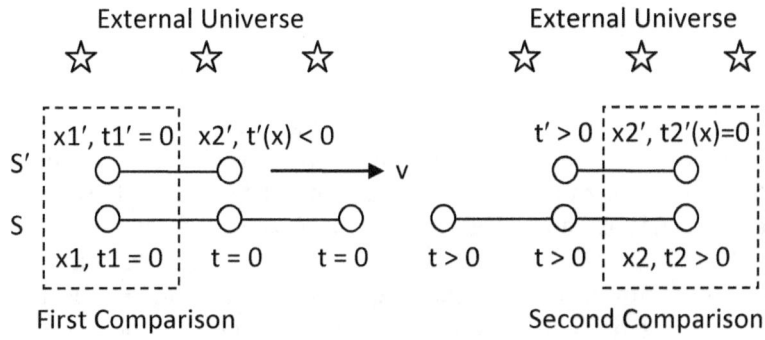

Fig.49: Apparent Covariant Length Contraction

Minkowski Diagrams

There are also implications for other areas of physics, such as Minkowski Diagrams. Devised by Hermann Minkowski after Einstein published his theory, the diagram was originally asymmetric, to show how a frame's space-time changed relative to another. His idea was later modified by others, giving versions that showed the symmetric (covariant) aspects of Special Relativity. In Machian Relativity, covariance is apparent, due to absolute relativistic simultaneity in the frame with absolute motion. The diagram can be re-drawn with the simultaneity (and covariance) <u>factored out</u>, to reveal the underlying asymmetry (e.g. the absolute motion of S' relative to S). The diagram is then returned to its original asymmetric form:

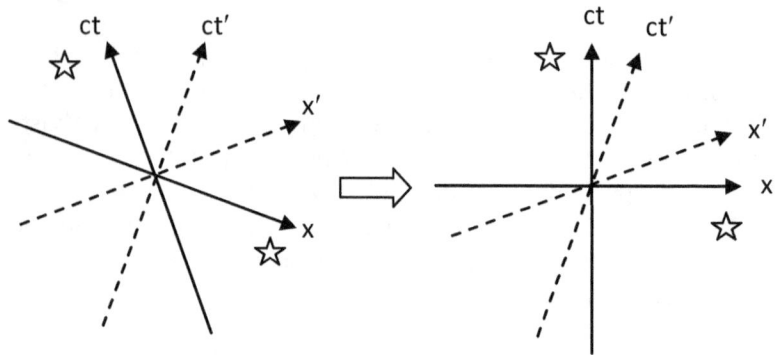

Fig.50: Factoring out Apparent Covariance

Their original purpose was purely illustrative. Nowadays however, such symmetric diagrams are often given as "proof" of Special Relativity. For example, "the covariance of Special Relativity is a fact because it can be represented by space-time rotations". But such diagrams only change an argument from an algebraic to a geometric form. That is, they represent the algebraic aspects of Special Relativity in graphical form and as such, they do not prove any particular argument. They only proper proof that can be given is by experimental evidence – if no experiment can specifically demonstrate the reciprocal aspect of Special Relativity, then the theory is false and there must be absolute motion and asymmetric time dilation. Therefore, an alternative theory is required which retains the c-invariance Postulate but abandons Einstein's Relativity Postulate as he envisioned it.

Factoring Out Apparent Covariance

The effects of relativistic simultaneity on covariance can be factored out in two ways. Firstly, in the Lorentz Transformations, the equations {3a} used to predict the behaviour of S (absolute rest) relative to S' (absolute motion) need to be treated in a different way from the other set {1a}. These equations are:

$$x = (x' + vt') / \gamma \qquad t = (t' + vx'/c^2) / \gamma \qquad \{3a\}$$

In the equation for calculating the "inverse time dilation" of S relative to S', two clocks in S must be compared to one clock in S' - not one clock in S compared to two clocks in S'. This factors out the absolute simultaneity effects in S', and $x1' = x2'$. Thus:

$$(t2 - t1) = (t2' - t1') / \gamma \qquad \{2b\}$$

For calculating the "inverse length contraction" of S relative to S, a rod in S' is directly compared to a rod in S when both origins coincide, even though the measurements made in S' (at each end of its rod) are not synchronous. Due to the absolute simultaneity of S', a clock on the end of the rod at the origin reads $t' = 0$ while the clock at the far end of the rod (along the x' axis) will read:

$$t2' = -vx2 / c^2 \sqrt{(1 - v^2/c^2)} = -vx2 / c^2\gamma$$
And: $\quad t1' = -vx1 / c^2 \sqrt{(1 - v^2/c^2)} = -vx1 / c^2\gamma$

Using these values in the Lorentz Equation for calculating x gives:

$$(x2 - x1) = (x2' + vt2' - x1' - vt1') / \gamma$$
$$= x2' / \gamma - v^2 x2 / c^2\gamma^2 - x1 / \gamma + v^2 x1 / c^2\gamma^2$$

$$(x2 - x1) + (x2 - x1) v^2 / c^2\gamma^2 \quad = (x2' - x1') / \gamma$$
$$(x2 - x1) / \gamma^2 \qquad\qquad = (x2' - x1') / \gamma$$

Thus: $\quad (x2 - x1) = (x2' - x1') \gamma \qquad \{2b\}$

So to summarise, the Lorentz Transforms predict the *relative measurements* made by each frame (after correcting for absolute Machian relativistic simultaneity in S') to be:

$$(x2' - x1') = (x2 - x1) / \gamma \qquad (t2' - t1') = (t2 - t1) \gamma \qquad \{1b\}$$
$$(x2 - x1) = (x2' - x1') \gamma \qquad (t2 - t1) = (t2' - t1') / \gamma \qquad \{2b\}$$

Thus, Machian Relativity predicts:

In S:	$L' = \gamma L$	$T' = \gamma T$	$\{1b\}$
In S':	$L = \gamma L'$	$T = \gamma T'$	$\{3b\}$ (Apparent)
In S':	$L = L' / \gamma$	$T = T' / \gamma$	$\{2b\}$ (True)

Alternatively, clocks in S' can be resynchronized by observers (in either S' or S) so that they are all synchronous with those in S. This removes the relativistic simultaneity in S' allowing observers in S' to measure the rate of a clock in S using two clocks in S'. This allows the inverse dilation of S to be directly measured, but the speed of light invariance along the axis of motion for S' is violated. However, there is still time dilation and length contraction. These maintain c-invariance in S' along the y'-axis and reflection along the x'-axis respectively. With simultaneity in S' factored out, the Lorentz Equations for S and S' are as follows:

$$x' = (x - v\,t) / \gamma \qquad\qquad t' = \gamma\,t \qquad \{1a\}$$
$$x = \gamma\,(x' + v'\,t') \qquad\qquad t = t' / \gamma \qquad \{2a\}$$

Where $v' = v / \gamma^2$.

Relativistic Four-Vectors

Relativistic four-vectors are treated in a similar way to that in Special Relativity. For example, four-momentum is given by:

$$\underline{P} = (E / c, \underline{p}) = (m_0 c / \sqrt{(1 - v^2/c^2)}, m_0\underline{v} / \sqrt{(1 - v^2/c^2)})$$
$$= m_0 (c, \underline{v}) / \sqrt{(1 - v^2/c^2)}$$

In Machian Relativity, $\underline{v}$ is the velocity vector representing the absolute velocity relative to distant matter, not the purely relative velocity per-se between any two inertial frames as would be expected for Special Relativity. In most applications (such as with relativistic particles), the earth is a good approximation to a state of absolute rest, due to its low absolute motion and so $\underline{v}$ can be assumed to be relative to the earth.

Also these equations, like those for mass and energy, are never described in a covariant form that would be expected for Special Relativity (e.g. time dilation). This is further indication that the symmetry of the theory does not exist.

APPENDIX VII: SPECIAL RELATIVITY AND THE PARADOXES OF RELATIVISTIC VELOCITY

Consider three coordinate systems moving as follows:

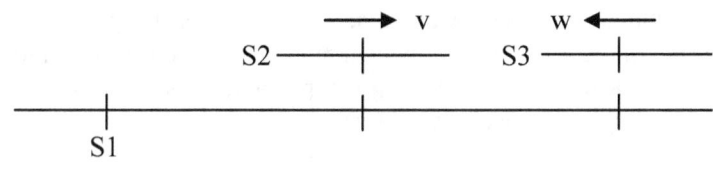

<u>Fig.51</u>

According to Special Relativity, if an inertial system S1 measures systems S2 and S3 to have speeds v and w respectively (in opposite directions so that S2 and S3 approach each other), then the relative velocity between S2 and S3 is:

$$u = (w + v) / (1 + w.v/c^2)$$

For relative speeds much lower than that of light, this equation approximates to Galilean Relativity. Thus, if $w.v/c^2 \approx 0$, then $u \approx (w + v)$. The equation also prevents u exceeding the speed of light. So, if (say) $v = w = 0.75c$, the equations gives:

$$u = 0.96c$$

Thus, the relative velocity is not 1.5c as expected by Galilean Relativity. In fact, if clocks in both S2 and S3 are synchronised according to the requirements of Special Relativity, it is impossible for the relative speed between these two systems to reach or exceed the speed of light.

Now, observers in these two systems can each measure the respective relative speed of S1, which is also their own speed relative to S1. They can add these two values (v and w) to get the Galilean combined speed and compare it to the directly measured value. The fact that observers in these two systems measure their combined relative velocity u to be different from Galilean Relativity (invariant space-time) proves that their systems have undergone physical changes in their space-times (clocks and rulers) due to their respective motions relative to S1. Therefore, the idea that the equations of Special Relativity are somehow "not physical" or "kinematic" because they only describe how a moving system "looks relative to an observer" is false. These equations describe objectively true physical changes, and this contradicts Einstein's Relativity Postulate, where observers in any inertial system consider themselves "at rest".

Corollary

Consider also the following example. Let the relative inertial speed between S1 and S3 be w. S2 is then introduced, which is also inertial and so "at rest" for its observers, and it moves at speed v = w/2 relative to S1, in the <u>same direction</u> as S3, so that it is always at the midpoint of S1 and S3 (x/2):

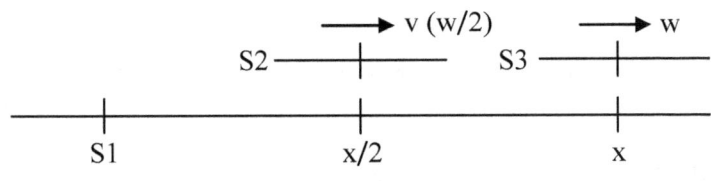

<u>Fig.52</u>

If S1 is "at rest" according to Special Relativity, the relative speed between observers in S2 and S3 is:

$$u = (w - v) / (1 - w.v/c^2) = (w / 2) / (1 - w^2 / 2c^2)$$
$$= w / (2 - w^2 / c^2) \neq w / 2$$

Thus $u \neq v$, even though S2 is at the midpoint of S1 and S3 during its motion.

Galilean Relativity predicts that the combined speed between S1 and S3 to be w = (u + v). However, the equation for the combined *relativistic* speed between S1 and S3 is:

$$W = (u + v) / (1 + u.v/c^2).$$

Because S2 is "at rest" for its observers, we would expect u = v = w / 2, because S2 is always at the midpoint. Thus, S1 and S3 recede equally from it at w / 2, and the combined relative velocity between S1 and S3 is:

$$W = (w/2 + w/2) / (1 + w^2/4c^2)$$
$$= w / (1 + w^2/4c^2) \neq w$$

Thus it can be seen that different results are generated depending on whether S1 or S2 is defined as being "at rest". If S1 is defined "at rest", then the speeds of S1 and S3 relative to S2 are unequal, even though S2 is at their mutual midpoint. Yet if S2 is "at rest", so that S1 and S3 have the same speed relative to S2, then their combined speed W is different from the value w when S1 was defined "at rest".

Relativistic Velocity in Machian Relativity

The above problems are resolved by Machian Relativity. If S1 is at rest relative to the External Universe, then S2 and S3 have *absolute* motion and undergo *physical* (asymmetric) changes:

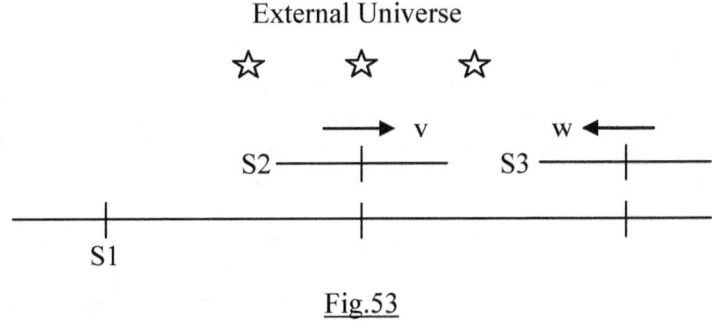

Fig.53

197

Thus the combined velocity between them is given by:

$$u = (w + v) / (1 + w.v/c^2)$$

Where: u is the combined *relative* velocity between S2 and S3, and w and v are the *absolute* velocities of S2 and S3 relative to the average momentum of all matter (corresponding to S1). The speeds w and v approximate to motion relative to the "Fixed Stars", and for large velocities this also approximates to motion relative to the earth. Because observers in S2 and S3 determine u < (w + v), this proves their clocks and rulers have undergone objective and physical changes, so their time dilations, length contractions and relativistic simultaneity change asymmetrically – even when they move inertially. Therefore, <u>Einstein's Relativity Postulate is false</u>[14].

The physical changes to the space-times of S2 and S3 are caused by motion relative to the average momentum of matter in the universe, and this explains the privileged status of S1. These changes maintain c-invariance for observers in S2 and S3, and prevent u reaching or exceeding the speed of light. Light from all sources, including their own, has absolute motion relative to S1 and the "Fixed Stars" (satisfying De Sitter's Binary Star Principle). According to Machian Relativity, such changes in S2 and S3 cause apparent symmetry in S1's changes. The same applies to the reciprocal nature of velocity (e.g. where observers in S1 and S2 measure the same relative speed for each other).

---

[14] Observers in S2 and S3 can perform the Michelson-Morley Experiment and get a null result, but this does not prove there is no absolute motion. Rather, the null result is due to a physical contraction of their rulers due to motion relative to S1, which is relative to the "Fixed Stars". Such physical changes are consistent with the Hafele-Keating Experiment, where both the aircraft clock and earth clock have asymmetric dilation relative to a non-rotating frame. Because earth observers have to include the effects of their own motion, this disproves Einstein's idea that they are "at rest". There is no reciprocal dilation between the earth and aircraft clocks.

APPENDIX VIII: STELLAR ABERRATION AND MACH'S PRINCIPLE[15]

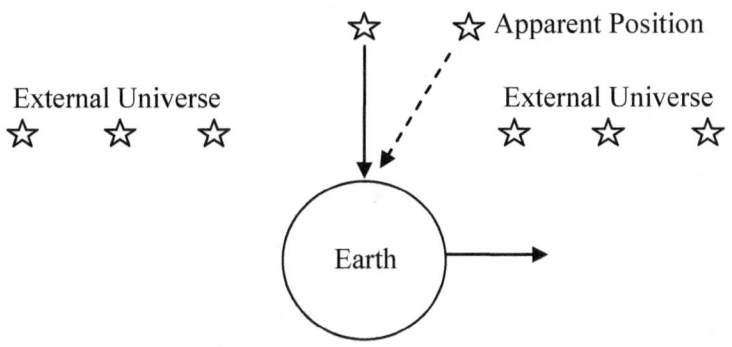

Fig.54a: Stellar Aberration due to Machian Relativity

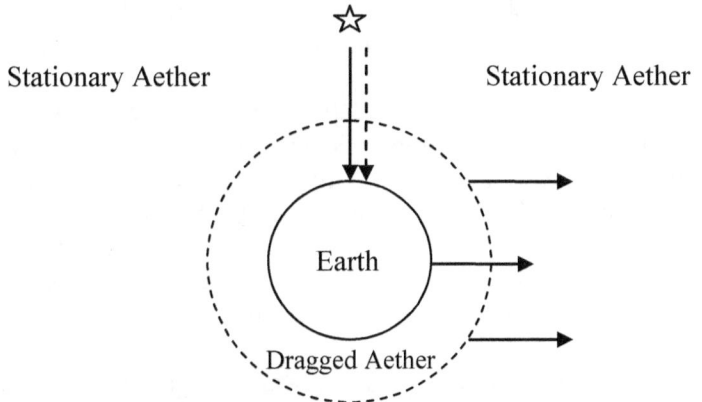

Fig.54b: Non-aberration due to Aether Drag

---

[15] Machian Relativity and Lorentz's Aether Relativity correctly predict stellar aberration. Aether Drag does not predict the effect so is false. Newtonian Relativity also predicts the effect, but is false because binary star observations prove there is an absolute frame for the motion of light. Stellar aberration is the earth's motion relative to starlight, which (because of this absolute frame) is the same motion relative to light from sources on earth. Hence, the frame of the earth is not at rest. This is consistent with the asymmetric time dilation of earth clocks in the Hafele-Experiment, and so Einstein's Relativity Postulate is false.

APPENDIX IX: THE INFLUENCE OF THE EXTERNAL UNIVERSE ON CLOCK OBSERVATIONS

Consider two coordinate systems in "relative" motion according to the Theory of Machian Relativity:

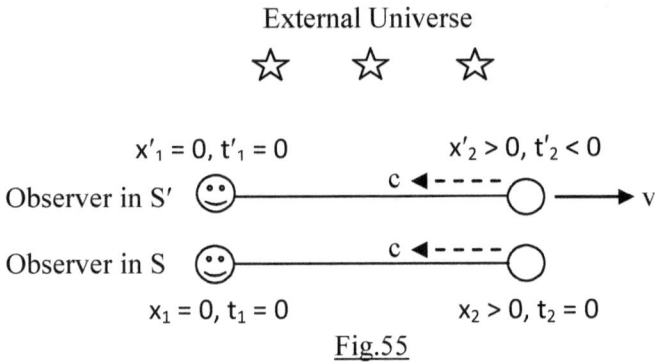

External Universe

Fig.55

Let S be at *absolute rest* relative to the External Universe according to Mach's Principle. S' moves relative to S and has *absolute motion*. Light from sources in S and S' has absolute motion relative to the External Universe (and S), thereby satisfying De Sitter's Principle. S and S' both measure the speed of light as constant; S because it is at absolute rest; and S' because it undergoes <u>physical changes</u> in its space-time.

The readings on clocks at x_2 and x'_2 are sent as light signals to observers at the origins of both systems (x_1 and x'_1). S's clock readings are all synchronised to read $t_1 = t_2 = 0$ when the origins of both systems coincide, while the readings of S''s clocks vary according to *relativistic simultaneity*: $t'_1 = 0$, $t'_2 = -vx_2/c^2\sqrt{(1-v^2/c^2)} = -vx_2/c^2\gamma$. The time for both light signals to reach S's observer is x_2/c, because all light has absolute motion relative to S. Thus when his clock reads $t_1 = 0$, he sees t_2 as it was $-x_2/c$ ago, and t'_2 as it was $-\gamma x_2/c$ ago (due to its time dilation). Thus the observer at *absolute* rest in S <u>sees</u>:

$$t_2 = 0 - x_2/c = -x_2/c$$
$$t'_2 = -vx_2/c^2\gamma - \gamma x_2/c$$

Glyn Phillips

Corollary: What the Moving Observer sees

The speed of S′ relative to S is v, so relative to S, the signal from either clock reaches S′′'s observer at $x = vt = x_2 - ct$. Thus:

$$vt + ct = x_2, \text{ or: } t = x_2 / (c+v)$$

Because of time dilation in S′, the corresponding reading for S′′'s observer is:

$$t' = \gamma t = \gamma x_2 / (c+v)$$

Thus S′′'s observer sees the clocks at x_2 and x'_2 read $t_2 = 0$ and t'_2 when his clock reads t′. So when his clock reading was $t' = 0$ (= t'_1), he would have seen x'_2's clock read $t'_2 = -vx_2/c^2\gamma - t'$. And because of S's (relative) *anti-dilation*, he correspondingly sees x_2's clock read $t_2 = 0 - t'/\gamma$. Also S′ has <u>absolute</u> Lorentz contraction, so its corresponding length measurements are <u>greater</u> than S's: $x' = x/\gamma$ (and $x = \gamma x'$). Thus he <u>sees</u>:

$$t_2 = 0 - t'/\gamma = -x_2 / (c+v) = -\gamma x'_2 / (c+v)$$
$$= vx'_2 / c^2\gamma - x'_2 / c\gamma$$

$$t'_2 = -vx_2/c^2\gamma - t' = -vx_2/c^2\gamma - \gamma x_2/(c+v) = -x_2/c\gamma$$
$$= -x'_2/c$$

Thus, at $t'_1 = 0$, the absolutely moving observer sees the clock at x'_2 read $-x'_2/c$, which is consistent with being "at rest". However, he sees the clock x_2 read $vx'_2/c^2\gamma - x'_2/c\gamma$, where we would instead expect $vx'_2/c^2\gamma - \gamma x'_2/c$ (if clocks in S were reciprocally dilated relative to S′ according to Special Relativity).

So there is an *asymmetry* in how these inertially moving observers see each other's clocks. Even if this reasoning is disputed for inertial observers, it <u>must still apply</u> when the systems are in "relative" motion, but one of them has a residual acceleration due to an applied force, if it is acknowledged that non-inertial effects cause asymmetries in Special Relativity.

www.ingramcontent.com/pod-product-compliance
Lightning Source LLC
Chambersburg PA
CBHW051456170526
45166CB00001B/267